Advances in Intelligent and Soft Computing 92

Editor-in-Chief: J. Kacprzyk

Advances in Intelligent and Soft Computing

Editor-in-Chief

Prof. Janusz Kacprzyk
Systems Research Institute
Polish Academy of Sciences
ul. Newelska 6
01-447 Warsaw
Poland
E-mail: kacprzyk@ibspan.waw.pl

Further volumes of this series can be found on our homepage: springer.com

Vol. 80. N.T. Nguyen, A. Zgrzywa, and A. Czyzewski (Eds.)
Advances in Multimedia and Network Information System Technologies, 2010
ISBN 978-3-642-14988-7

Vol. 81. J. Düh, H. Hufnagl, E. Juritsch, R. Pfliegl, H.-K. Schimany, and Hans Schönegger (Eds.)
Data and Mobility, 2010
ISBN 978-3-642-15502-4

Vol. 82. B.-Y. Cao, G.-J. Wang, S.-L. Chen, and S.-Z. Guo (Eds.)
Quantitative Logic and Soft Computing 2010
ISBN 978-3-642-15659-5

Vol. 83. J. Angeles, B. Boulet, J.J. Clark, J. Kovecses, and K. Siddiqi (Eds.)
Brain, Body and Machine, 2010
ISBN 978-3-642-16258-9

Vol. 84. Ryszard S. Choraś (Ed.)
Image Processing and Communications Challenges 2, 2010
ISBN 978-3-642-16294-7

Vol. 85. Á. Herrero, E. Corchado, C. Redondo, and Á. Alonso (Eds.)
Computational Intelligence in Security for Information Systems 2010
ISBN 978-3-642-16625-9

Vol. 86. E. Mugellini, P.S. Szczepaniak, M.C. Pettenati, and M. Sokhn (Eds.)
Advances in Intelligent Web Mastering – 3, 2011
ISBN 978-3-642-18028-6

Vol. 87. E. Corchado, V. Snášel, J. Sedano, A.E. Hassanien, J.L. Calvo, and D. Ślęzak (Eds.)
Soft Computing Models in Industrial and Environmental Applications, 6th International Workshop SOCO 2011
ISBN 978-3-642-19643-0

Vol. 88. Y. Demazeau, M. Pěchouček, J.M. Corchado, and J.B. Pérez (Eds.)
Advances on Practical Applications of Agents and Multiagent Systems, 2011
ISBN 978-3-642-19874-8

Vol. 89. J.B. Pérez, J.M. Corchado, M.N. Moreno, V. Julián, P. Mathieu, J. Canada-Bago, A. Ortega, and A.F. Caballero (Eds.)
Highlights in Practical Applications of Agents and Multiagent Systems, 2011
ISBN 978-3-642-19916-5

Vol. 90. J.M. Corchado, J.B. Pérez, K. Hallenborg, P. Golinska, and R. Corchuelo (Eds.)
Trends in Practical Applications of Agents and Multiagent Systems, 2011
ISBN 978-3-642-19930-1

Vol. 91. A. Abraham, J.M. Corchado, S.R. González, J.F. de Paz Santana (Eds.)
International Symposium on Distributed Computing and Artificial Intelligence, 2011
ISBN 978-3-642-19933-2

Vol. 92. P. Novais, D. Preuveneers, and J.M. Corchado (Eds.)
Ambient Intelligence - Software and Applications, 2011
ISBN 978-3-642-19936-3

Paulo Novais, Davy Preuveneers, and
Juan M. Corchado (Eds.)

Ambient Intelligence – Software and Applications

2nd International Symposium on Ambient Intelligence (ISAmI 2011)

Springer

Editors

Prof. Paulo Novais
Universidade do Minho
Departamento de Informática
Campus de Gualtar
4710-057 Braga
Portugal

Dr. Davy Preuveneers
Katholieke Universiteit Leuven
Department of Computer Science
Celestijnenlaan 200A, 02.152
B-3001 Heverlee
Belgium

Prof. Juan M. Corchado
Universidad de Salamanca
Departamento de Informática
y Automática
Facultad de Ciencias
Plaza de la Merced S/N
37008, Salamanca
Spain
E-mail: corchado@usal.es

ISBN 978-3-642-19936-3 e-ISBN 978-3-642-19937-0

DOI 10.1007/978-3-642-19937-0

Advances in Intelligent and Soft Computing ISSN 1867-5662

Library of Congress Control Number: 2011923221

Typeset & Cover Design: Scientific Publishing Services Pvt. Ltd., Chennai, India.

Printed on acid-free paper
5 4 3 2 1 0
springer.com

Preface

Ambient Intelligence (AmI) emerged about a decade ago as a user-centred computing paradigm envisioning the information society of 2010-2020. In AmI, the focus is on an almost invisible integration of computational power and communication technologies in everyday objects, ranging from smartphones to consumer electronics, to guarantee a natural and user friendly human computer interaction experience. Nowadays, this inter-disciplinary field is evolving quickly due to new challenges posed by innovative applications in the area of health and wellness, smart environments for assisted living, social networks, ambient media and entertainment, intelligent transportation, and other personalized services that aim to make everyone's life even more comfortable.

For AmI to flourish, interaction with technology in the surroundings should be smooth and happen without people actually noticing it. The only awareness people should have arises from the goals of AmI itself: more safety, comfort and wellbeing, emerging in a natural way.

ISAmI is the International Symposium on Ambient Intelligence, and aims to bring together researchers from various disciplines that are interested in all aspects of Ambient Intelligence. The symposium provides a forum to present and discuss the latest results, innovative projects, new ideas and research directions, and to review current trends in this area.

This volume presents the papers that have been accepted for the 2011 edition, both for the main event and workshop. The ISAmI workshop WoRIE promises to be a very interesting event that complements the regular program with an emerging topic on reliability of intelligent environments. The papers in these proceedings report on innovative results and advances achieved recently in AmI area. With a full programme composed of 22 long papers and 7 short papers, and 4 long papers accepted in the workshop, these proceedings capture the most innovative results and advances in 2011. Each paper has been reviewed by, at least, three different reviewers, from an international committee composed of 44 members from 23 countries.

We would like to thank all the contributing authors, as well as the members of the Program Committee and the Local Organizing Committee for their hard and highly valuable work. Your work has helped the ISAmI symposium to become a success. Thanks for your help; ISAmI 2011 would not have existed without your contribution.

April 2011

The editors
Paulo Novais
Davy Preuveneers
Juan M. Corchado

ISAmI'11 has been supported by Junta de Castilla y León (Spain).

Organization

Scientific Committee Chairs

Paulo Novais	University of Minho (Portugal)
Davy Preuveneers	Katholieke Universiteit Leuven (Belgium)

Organizing Committee Chair

Juan M. Corchado	University of Salamanca (Spain)

Steering Committee

Juan Carlos Augusto	University of Ulster (UK)
Juan M. Corchado	University of Salamanca (Spain)
Paulo Novais	University of Minho (Portugal)

Program Committee

Aaron Quigley	Human Interface Technology Laboratory (HITLab) (Australia)
Alireza Sahami	Univeristy of Duisburg Essen (Germany)
Andreas Riener	Johannes Kepler University Linz (Austria)
Antonio F. Gómez Skarmeta	University of Murcia (Spain)
Carlos Juiz	University of the Balearic Islands (Spain)
Carlos Ramos	Polytechnic of Porto (Portugal)
Cecilio Angulo	Polytechnic University of Catalonia (Spain)
Celia Gutierrez	Complutense University of Madrid (Spain)
Cristina Buiza	Ingema (Spain)
Diego Gachet	European University of Madrid (Spain)
Eduardo Dias	University of Évora (Portugal)
Elisabeth Eichhorn	Potsdam University of Applied Sciences (Germany)
Emilio S. Corchado	University of Burgos (Spain)
Francisco C. Pereira	University of Coimbra (Portugal)
Francisco J. Ballesteros	Rey Juan Carlos University (Spain)

Goreti Marreiros	Polytechnic of Porto (Portugal)
Gregor Broll	DOCOMO Euro-Labs (Germany)
Guillaume Lopez	University of Tokyo (Japan)
Hani Hagras	University of Essex (UK)
Javier Jaen	Polytechnic University of Valencia (Spain)
Jo Vermeulen	Hasselt University (Belgium)
José C. Danado	NTNU (Norway)
José M. Molina	University Carlos III of Madrid (Spain)
José Machado	University of Minho (Portugal)
Joyca Lacroix	Philips Research (Netherlands)
Kristof Van Laerhoven	TU Darmstadt (Germany)
Lourdes Borrajo	University of Vigo (Spain)
Martijn Vastenburg	Delft University of Technology (Netherlands)
Rene Meier	Trinity College Dublin (Ireland)
Rui José	University of Minho (Portugal)
Teresa Romão	New University of Lisbon (Portugal)
Wolfgang Reitberger	University of Salzburg (Austria)
Simon Egerton	Monash University (Malaysia)
Usa Sammapun	Kasetsart University (Thailand)
Jaderick Pabico	University of the Philippines Los Baños (Philippines)
Junzhong Gu	East China Normal University (China)
George Demiris	University of Washington (USA)
Alex Mihailidis	University of Toronto (Canada)
Diane Cook	Washington State University (USA)
Hans Guesgen	Massey University (New Zealand)
Flavia Delicato	Federal University of Rio Grande do Norte (Brasil)
Tatsuya Yamazaki	NICT (Japan)
Yu Zheng	Microsoft Research (Asia)
Veikko Ikonen	VTT Technical Research Centre (Finland)

Workshop on the Reliability of Intelligent Environments

Pablo A. Haya Coll	Autonomous University of Madrid (Spain)
Juan Carlos Augusto	University of Ulster (UK)

Local Organization Committee

Juan M. Corchado	University of Salamanca (Spain)
Dante I. Tapia	University of Salamanca (Spain)
Javier Bajo	Pontifical University of Salamanca (Spain)
Fernando de la Prieta	University of Salamanca (Spain)
Cesar Analide	University of Minho (Portugal)

Contents

Long Papers

AmI Support for the Trading Process: Self-aware Trader Model 1
Javier Martínez Fernández, Juan Carlos Augusto, Ralf Seepold, Natividad Martínez Madrid

Design Time Methodology for the Formal Verification of Intelligent Domotic Environments 9
Fulvio Corno, Muhammad Sanaullah

Wheelchair-Based System Adapted to Disabled People with Very Low Mobility 17
Albano Carrera, Alonso A. Alonso, Ramón de la Rosa Steinz, María I. Jiménez Gómez, Lara del Val

Toward Seamless Environments for Dispute Prevention and Resolution 25
Davide Carneiro, Paulo Novais, José Neves

Integrating Personalized Health Care and Information Access for Elder People 33
Diego Gachet Páez, Juan R. Ascanio, Ignacio Giráldez, Margarita Rubio

A DSL for Context Quality Modeling in Context-Aware Applications 41
José R. Hoyos, Davy Preuveneers, Jesús J. García-Molina, Yolande Berbers

Non-intrusive Residential Electrical Consumption Traces 51
Marisa Figueiredo, Ana de Almeida, Bernardete Ribeiro

Social Presence in Immersive 3D Virtual Learning Environments 59
Minjuan Wang, Sabine Lawless-Reljic, Marc Davies, Victor Callaghan

A Digital Secretary for Smart Offices Setup Up 69
João Laranjeira, Carlos Freitas, Goreti Marreiros, Carlos Ramos, João Carneiro

Applying HoCCAC to Plan Task the COPD Patient: A Case Study 77
Juan A. Fraile, Sara Rodríguez, Juan F. de Paz, Belén Pérez-Lancho

Elder Care's Fall Detection System 85
Filipe Felisberto, Nuno Moreira, Isabel Marcelino, Florentino Fdez-Riverola, and António Pereira

Multi-Agent System for Detecting Elderly People Falls Through Mobile Devices 93
Patricia Martín, Miguel Sánchez, Laura Álvarez, Vidal Alonso, Javier Bajo

Intelligent Video Monitoring for Anomalous Event Detection 101
Iván Gómez Conde, David Olivieri Cecchi, Xosé Antón Vila Sobrino, Ángel Orosa Rodríguez

Design and Modelling of the Nocturnal AAL Care System ... 109
J.C. Augusto, H. Zheng, M. Mulvenna, H. Wang, W. Carswell, P. Jeffers

The EducAgent Platform: Intelligent Conversational Agents for E-Learning Applications 117
David Griol, Jesús García-Herrero, José M. Molina

Improving Human Face Detection through TOF Cameras for Ambient Intelligence Applications 125
J.R. Ruiz-Sarmiento, C. Galindo, J. Gonzalez

Recommendation and Planning through Mobile Devices in Tourism Context 133
Ricardo Anacleto, Lino Figueiredo, Nuno Luz, Ana Almeida, Paulo Novais

Mobile Personal Health Systems for Patient Self-management: On Pervasive Information Logging and Sharing within Social Networks 141
Andreas K. Triantafyllidis, Vassilis G. Koutkias, Ioanna Chouvarda, Nicos Maglaveras

Multi-Agent Strategy Synthesis in Smart Meeting Environments 149
René Leistikow

Reusable Gestures for Interacting with Ambient Displays in Unfamiliar Environments 157
Radu-Daniel Vatavu

Graphs and Patterns for Context-Awareness 165
Andrei Olaru, Adina Magda Florea, Amal El Fallah Seghrouchni

ARTIZT: Applying Ambient Intelligence to a Museum Guide Scenario 173
Oscar García, Ricardo S. Alonso, Fabio Guevara, David Sancho, Miguel Sánchez, Javier Bajo

Workshop on the Reliability of Intelligent Environments (WoRIE)

Reliability of Location Detection in Intelligent Environments 181
Shumei Zhang, Paul J. McCullagh, Chris Nugent, Huiru Zheng, Norman Black

Flexible Simulation of Ubiquitous Computing Environments 189
Francisco Campuzano, Teresa Garcia-Valverde, Alberto Garcia-Sola, Juan A. Botia

Safety Considerations in the Development of Intelligent Environments 197
Juan Carlos Augusto, Paul J. McCullagh

LECOMP: Low Energy COnsumption Mesh Protocol in WSN 205
Juan A. Ortega, Alejandro Fernandez-Montes, Daniel Fuentes, Luis Gonzalez-Abril

Short Papers

A User-Friendly Interface for Rules Composition in Intelligent Environments ... 213
Dario Bonino, Fulvio Corno, Luigi De Russis

Using an Ambient Intelligent Architecture for Developing an Intelligent Tutoring System for Training Operators of Power System Control Centres ... 219
Luiz Faria, António Silva, Carlos Ramos, Luís Gomes, Zita Vale, Albino Marques

Ambient Intelligence Based Architecture for Immersive Social Entertainment ... 227
Mª Amparo Navarro, Ana Belén Sánchez, Carlos Fernández, Mª Teresa Meneu

Grouping Behaviour in AmI-Enabled Crowd Evacuation ... 233
Alexei Sharpanskykh, Kashif Zia

A Mobile System to Visualize Patterns of Everyday Life ... 241
Nuno Correia, Armanda Rodrigues, Tiago Amorim, Jared Hawkey, Sofia Oliveira

Human Behaviour Modelling Approach for Intention Recognition in Ambient Assisted Living ... 247
Kristina Yordanova

Evolutionary Intelligence in Agent Modeling and Interoperability ... 253
Miguel Miranda, José Machado, António Abelha, José Neves, João Neves

Author Index ... 259

AmI Support for the Trading Process: Self-aware Trader Model

Javier Martínez Fernández, Juan Carlos Augusto,
Ralf Seepold, and Natividad Martínez Madrid

Abstract. Decision making in trading can be compromised when traders are under stress. In this paper is presented a case study where ambient intelligence has been used to provide traders with self-awareness of their stress levels. We illustrate how using specific stress sensors improved the results of the trading process by avoiding bad decision making at crucial moments.

Keywords: AmI, Stress Awareness, Trading Process, Stressful Decision Making.

1 Introduction

Decision making in the trading process is a clear example of an activity where risk management in real time is important. Risks have a high impact in decision making because they can undermine our capability to make safe decisions. Nowadays there are several tools to help traders in their daily operative and although this is really good and helpful, the major risk source is missing, the trader's state-of-mind self-awareness. A trader who is self-aware of her/his own stress can have a more effective and coherent decision making.

Javier Martínez Fernández
University Carlos III of Madrid, Spain
e-mail: `javier.martinez@uc3m.es`

Juan Carlos Augusto
University of Ulster, United Kingdom
e-mail: `jc.augusto@ulster.ac.uk`

Ralf Seepold
University of Applied Sciences Konstanz, Germany
e-mail: `ralf.seepold@htwg-konstanz.de`

Natividad Martínez Madrid
Reutlingen University, Germany
e-mail: `natividad.martinez@reutlingen-university.de`

P. Novais et al. (Eds.): Ambient Intelligence - Software and Applications, AISC 92, pp. 1–8.
springerlink.com

Ambient Intelligence as an emerging area focused on building digital environments that proactively, but sensibly, support people in their daily lives [1] is relevant to the problem. Technical studies [2] have proven that stress produces biometric changes in the human body and in the effectiveness of the decision making. These changes can be measured with AmI support by means of sensors to bring self-awareness to the trader. We illustrate our system in this paper through a case study showing how Self-Aware traders improve their performance and results. Next section explains how decision making is impaired under stress. Section three shows the Self-Aware trader model. The case study performed is described in section four. Finally, conclusions and future work are presented.

2 Motivations for a Trader

Good judgment during decision making can be severely impaired by stress [3]. The most accepted theory is that under stress we scan fewer alternatives searching the solution of a problem [4]. Selten et al [5] explain that when we do decision-making we use a toolbox of strategies and we apply the strategy with the most adaptive heuristic available.

Decision making performed during trading is in real time and good timing is also crucial. This relates to stress because the search of alternatives is truncated by time limitations [6]. In addition to time limitations, decisions made during trading are inter-related and there is evidence that individuals tend to rely on previous responses regardless of their success [7], for this reason when a trader has a lost due to a wrong decision that can negatively affect the next decision.

All this body of research supports that decision making is degraded under stressful conditions. In particular, for the trading profession, detecting the moment in which the trader starts to become overwhelmed by stress is very important to avoid possible mistakes in the operations. The problem is that when the trader is under stress, the trader himself does not realize that (In psychological terms this is named illusion of control). However, given that stress also has a biological manifestation; it can be measured with Ami support through current state of the art sensing technology [8].

In this research line, Lo and Repin [9] reported with some experiments that physiological variables associated with the autonomic nervous system are highly correlated with market events. We continue in this direction with one more step. Helping traders to understand their mental state in real time as part of the trading information (news, charts, clients' accounts balance) can lead to safer decision making. In the next section we apply this concept in the Self-Aware Trader model.

3 Self-aware Trader Model

According to [10] when you're experiencing stressful emotions, whether you're conscious of them or not, higher brain processes become seriously compromised.

This phenomenon is called *cortical inhibition*. In the same study this cortical inhibition is tested with the impact of stress on the cardiovascular system in real time. The more stable the frequency and shape of the waveform of the heart rate, the more coherent the system becomes. In physiological terms, coherence describes the degree to which respiration and heart rate oscillate at the same frequency. The HeartMarth's sensor with Emwave software allows measuring this coherence/stress level in real time with only one sensor in the ear and the data can be easily shown in real time. It is an USB Plethysmographic pulse sensor for ear, optionally for finger with a sample rate of 360 samples/sec. The gain (increase needed for the amplitude of the signal) setting adjusts automatically via LED duty cycle (ratio between the pulse duration and the period of a rectangular waveform). The photo diode gain adjustment and the operating range is 30 - 140 beats/sec.

Guided by previous considerations [11] we arrive to a system model (Fig. 1) with the following features:

- The stress/coherence level will be processed in real time with the biometric wearable and not invasive Heartmarth sensor.
- An alert will only be shown to the trader when there is an indication that her/his stress level jeopardizes decision-making. This alert will disappear when the state of the trader comes back to a normal stress/coherence level, otherwise, the message "coherence/stress level in risk" is shown.

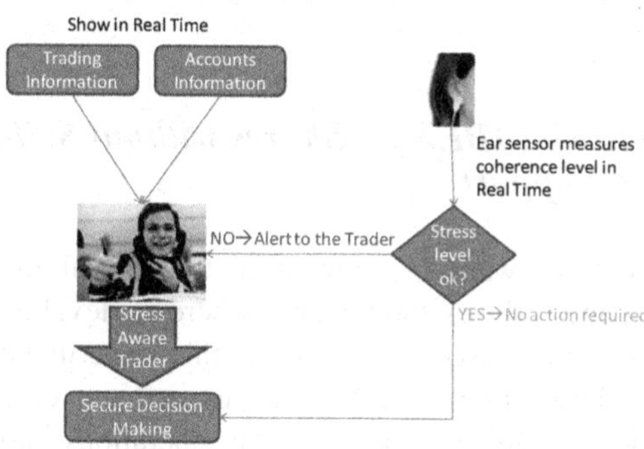

Fig. 1 Self-Aware Trader Model

The way in which the trader manages this information will depend on the context of the trader. If the trader is a self-employed person the action of stopping or continuing with the operation could be managed by him/herself. However, if the context is an investment bank, someone else can decide to stop the operative.

In both cases, the data sent through the sensor support alerts to the trader in real-time in an unobtrusive way. This is aligned with fundamental AmI principles. Independently if the operative is stopped or not, now we have a self-aware trader, therefore we have a safer decision making. Next section shows a case study with this model.

4 Case Study

The case study consists of a trader working in the financial market from a computer through internet (this is the most common case for traders in US). The stress sensor is connected in the trader's ear and we depict some significant moments of the trading session and how the sensed data and increased self-awareness influence the results of the trading exercise.

4.1 Scenario Setup

The equipment used by the trader is a laptop with internet connection, the HeartMath stress relief system (stress sensor) developed and manufactured by Quantum Intech, Inc. and the Emwave Pc (V1.0) software to see the stress level. The experiments have been done in the Spanish financial market, and all necessary information for the trading process (price of shares, charts, news...) has been extracted from Infobolsa web (http://www.infobolsa.es/). The trader who deploys the experiment is a non professional trader, but has 8 years of experience in this market trading for himself. To avoid problems with the income tax, commissions, and other effects of the trading process, the experiments have been done without real money. However, decision making is performed in real time in the real market and the aim to beat the market and win (play) money, provides motivation and a stress source in our trader.

4.2 Experiments with IBEX 35 Shares without Self-aware Trader Information Support

With this experiment we want to know if in the crucial moments of trading with shares (buying and selling), the trader's coherence level is affected and how that changes the decision making process. In this experiment, the Self-Aware information is not shown to the trader. The experiment duration is 1hour 30 minutes. In the next table, we can see the operations done by the trader (21/04/2010):

Table 1 Share Trading

Buy Time	Share	Price	Sell Time	Price
12:00	TELEFONICA	17.380	13:23	17.390
12:05	BBVA	10.970	13:22	10.980
12:06	SANTANDER	10.305	13:14	10.32
12:20	ABENGOA	20.10	13:04	20.27

Fig.2 shows the trader's coherence during trading time. The time is represented on axis X. The coherence level, as the Emwave program names it, Accumulated Coherence Score is shown in axis Y. When the coherence is in safe mode the score grows up, when the coherence enter in an unsafe mode, it decreases.

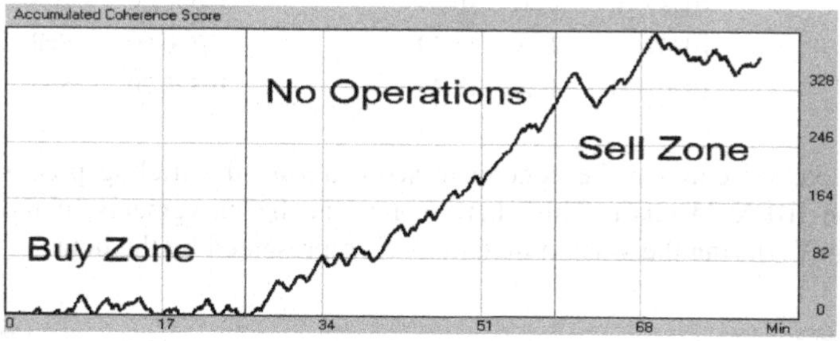

Fig. 2 Coherence Score in Session 1

We can see that the first 20 minutes of buying operations the stress of the trader prevents achieving good coherence levels (the lines in the graphic remains low). Later on, next half an hour of the experiment shows the time when the trader was waiting for the optimal selling point and that inactivity allows a high level of coherence to be reached, which is reflected on the graphic growing up. Finally, the stress again comes back when the trader tries to sell the shares at an optimal prize. The conclusion of the experiment is that the trader's body and coherence changes naturally reflect the crucial moments (buying and selling of shares).

4.3 Experiments with IBEX 35 Index Futures without Self-aware Information Support

In this other experiment, the trader operates with "futures" in IBEX-35 Index. The Index is an indicator formed by the principal companies of the national market, in US for example is Down Jones Index, in Spain it is formed by the 35 more important national companies and it is the IBEX-35. This index has a prize and moves it depending on the movement of the companies that component it. If the trader thinks that this index will go up, then the trader can buy an IBEX-35 "future" and sell it when the trader thinks that the index will go down. In the opposite case then trader can sell a "future" and buy it when s/he thinks the market could go up. In this operative each point up or down represents significant money and the movements are very fast (unlike in the shares trading). It forces traders to continue operations and increases the stress moments. Table 2 indicates the operations most representative for the experiment in 1 hour 30 minutes (22/04/2010). Taking into account that the first 20 minutes the trader considers not to make any operation, but the Emwave software is running and recording the coherence level. In this experiment, the information for Self-Awareness is not shown to the trader.

Table 2 Future Trading without Support

Open Time	Prize Open	OP	Close Time	Prize Close	OP	Points
16:21	10808.30	Buy	16:26	10830.20	Sell	+21,9
16:27	10843.10	Sell	16:31	10848.30	Buy	- 5,2
16:32	10849.90	Buy	16:39	10816.60	Sell	-33,3
16:45	10801.00	Buy	16:49	10828.20	Sell	+27,2

In Fig.3a, we can see the coherence score during the trading process and in Fig.3b the IBEX-35 chart. This chart represents the movements of the Index's prize (axis Y) during the session time (axis X) represented by the hour.

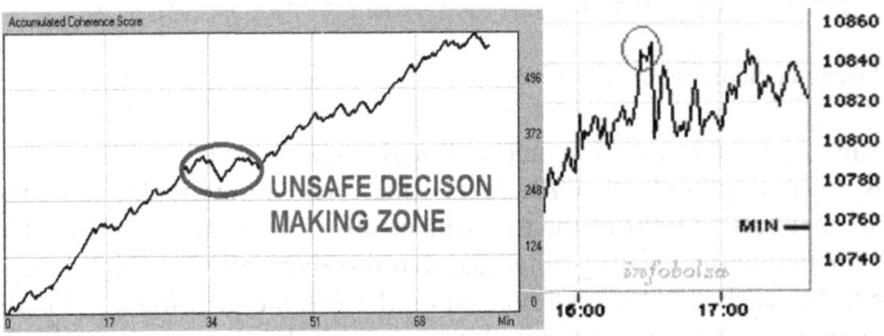

Fig. 3 Coherence Score in Session 2 (a) and Chart of IBEX-35 in session 2 (b)

We can see a frequent trading risk situation, at 16:27. The trader believes that 10843 is a good level to sell (corresponding with the first top of the graphic), then the trader sells. However, IBEX go up some points breaking the trader strategy, so the trader buys losing some points. The trader tries immediately to change of strategy, he knows that he was wrong and do not want to lose this crucial moment, now the trader is very stressed and his decision making is unsafe. The market has done a "false break" (circle in red in Fig 3.b) and the trader has bought in a rushed decision (circle of Fig 3.a). The trader loses 37 points in total. The Self-Aware information could have helped to avoid this mistake.

4.4 Experiments with IBEX 35 Index Futures with Self-aware Information Support

This experiment is conducted to show the favorable impact in the trading process when the trader has access to the self-aware information support. This session was the most difficult session for trading during the experiments, due to the Greece debt news at the time of the exercise. The Index suffered abrupt variations in seconds. Table 3 indicates the operations in 1 hour (23/04/2010). In this table, the number of sequence of operation has been added for a better tracking.

Table 3 Future Trading with Support

Open Time	Value Open	OP/nº	Close Time	Value Close	OP/nº	Points
16:25	10948.60	Sell/1	16:28	10948.60	Buy/2	0
16:31	10947.30	Buy/3	16:44	10952.80	Sell/4	+5,5
16:59	10921.70	Buy/5	17:01	10934.20	Sell/6	+12,5
17:03	10906.20	Buy/7	17:07	10923.70	Sell/8	+17,5

In Fig. 4 a) we can see the coherence score; in this case we have added numbers corresponding just with the 8 moments when trader made a decision to buy or sell. Fig. 4 b) shows the IBEX-35 chart.

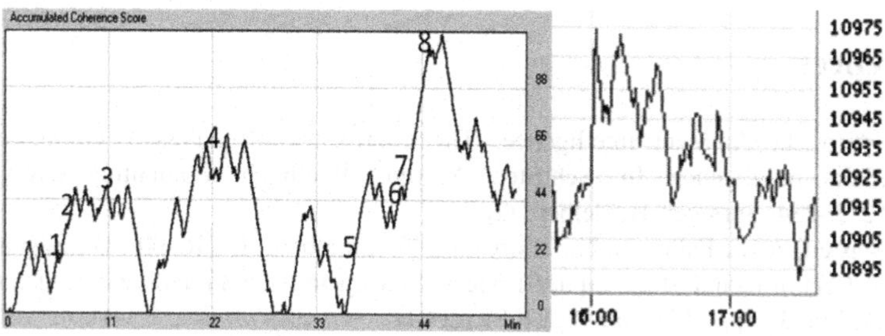

Fig. 4 Coherence Score in Session 3 (a) and Chart of IBEX-35 in session 3 (b)

In this session, we can see the high impact of the news and the abrupt movements of the Index in the coherence of the trader. However, in this case, the trader has access to feedback from the system on his coherence levels and based on that the trader decides to make decisions only when he believes to be in a safe state avoiding, in this case, bad operations.

After the experiments, we achieved the following conclusions: a) Previous losses, fear to lose a great trading movement, news, time restrictions, and many other factors have high impact in the trader's decision making; b) The more difficult the market is, the more important becomes for a trader to be Self-Aware of her/his level of stress; c) The greatest losses are the result of a wrong decision made at crucial time where being Self-Aware is very important. d) Traders with AmI support, Self-Aware Traders, improve significantly their results.

5 Conclusions and Future Work

A trader's mental state could be the most important information to know during trading. Currently this information is not usually available. We have conducted experiments to illustrate how to reduce the chances to fall when stress can adversely influence decision making in trading. Technological support based on sensors that facilitate context-awareness can give traders a Self-Awareness that is useful to avoid mistakes at crucial moments. Context-awareness in this case is achieved by feeding back to the trader their perception of the world of finances. The first experiment with shares, where the movement in the market was slower shows less impact. However, in the following experiments where the movements were very quicker and the stress was higher the advantages are more noticeable: decision making for Self-Aware traders is safer. Our future research will be focused on extending the Self-Aware Trader concept to a group of traders creating Group-Aware information. Besides, we will implement a system where this AmI support can be better integrated with the trading process.

Acknowledgments. This work has been partly funded by the Spanish Ministry of Industry under the project OSAMI (TSI-020400-2008-114), and the Spanish Ministry of Education under the project ARTEMISA (TIN2009-14378-C02-02).

References

1. Augusto, J.: Ambient intelligence: The confluence of pervasive computing and artificial intelligence. In: Schuster, A. (ed.) Intelligent Computing Everywhere, pp. 213–234. Springer, Heidelberg (2007)
2. Blangsted, A.K., Fallentin, N., Hjortskov, N., Lundberg, U., Rissén, D., Søgaard, K.: The effect of mental stress on heart rate variability and blood pressure during computer work. Eur. J. Appl. Physiol. 92, 84–89 (2004)
3. Delgado, M.R., Porcelli, A.J.: Acute Stress Modulates Risk Taking in Financial Decision Making. Psychol. Sci. 20(3), 278–283 (2009)
4. Speed, A., Forsythe, J.C.: Human emulation technology to aid the warfighter: Advances in computational augmentation of human cognition. White Paper (2002)
5. Gigerenzer, G., Selten, R.: Bounded rationality: The adaptive toolbox. MIT press, Cambridge (2001)
6. Dougherty, M.R.P., Hunter, J.: Probability judgment and subadditivity: The role of working memory capacity and constraining retrieval. Memory & Cognition 31(6), 968–982 (2003)
7. Lehner, P., Seyed-Solorforough, M., O'Connor, M.F., Sak, S., Mullin, T.: Cognitive biases and time stress in team decision making. IEEE Transactions on Systems, Man, & Cybernetics Part A: Systems & Humans 27, 698–703 (1997)
8. Martínez Fernández, J., Augusto, J., Seepold, R., Martínez Madrid, N.: Why Traders Need Ambient Intelligence. In: Augusto, J.C., Corchado, J.M., Novais, P., Analide, C. (eds.) ISAmI 2010. AISC, vol. 72, pp. 229–236. Springer, Heidelberg (2010)
9. Lo, A.W., Repin, D.V.: The Psychophysiology of Real-Time Financial Risk Processing, NBER Working Papers 8508, National Bureau of Economic Research, Inc. (2001)
10. Cryer, B., McCraty, R., Childre, D.: Pull the plug on stress. Harvard Business Review (2003)
11. Martínez Fernández, J., Augusto, J., Seepold, R., Martínez Madrid, N.: Sensors in Trading Process: A Stress–Aware Trader. In: 8th Proc. IEEE Workshop on Intelligent Solutions in Embedded Systems, pp. 17–21 (2010)

Design Time Methodology for the Formal Verification of Intelligent Domotic Environments

Fulvio Corno and Muhammad Sanaullah

Abstract. Ambient Intelligence systems integrate domotic devices and advanced control and intelligent algorithms, thus leading to integrate systems with a high degree of complexity in their behavior. Ensuring the correctness of the design of such system is therefore essential, and this paper proposed a methodology, based on formal modeling and verification techniques, to verify logic and temporal properties of an intelligent ambient. The approach is integrated with the Dog domotic gateway, and automatic translation tools ensure the correctness of the verified model while adding no additional task for system designers.

Keywords: Ambient Intelligence, Model Checking, Formal Verification, Statecharts.

1 Introduction

Ambient Intelligence is a promising and quickly expanding research field and is expected in the near future to have a wide impact over people's lifestyles [8]. Major electronic companies and various research and development organizations are working on the production and development of these systems [1]. In this context, current technology (often called *domotics*) already enables today's homes to perform automatic and intelligent behaviors, by means of networked electrical devices (sensors, actuators, consumer devices and various other controllable devices) in the home environment, governed by suitable algorithms, which work intelligently by also considering the presence and actions of people. Such algorithms are often implemented in an embedded computer called *home gateway*. We call this currently available home intelligence solutions: Intelligent Domotic Environments (IDEs) [2].

Fulvio Corno · Muhammad Sanaullah
Politecnico di Torino, Torino, Italy
e-mail: {fulvio.corno,muhammad.sanaullah}@polito.it

P. Novais et al. (Eds.): Ambient Intelligence - Software and Applications, AISC 92, pp. 9–16.
springerlink.com

Correct design of these IDEs is a challenge, since the interactions of many interconnected intelligent devices (composing the domotic systems) with the governing algorithms running in the home gateway may create complex global behaviors, that are difficult to predict before actually building and testing the system. System designers need powerful methodologies and tools to enable them to ensure the correctness of the system in early stages of the design process.

Some efforts to enable design-time validation propose the adoption of simulation based approaches, where a model of the system is expressed in an operational formalism that allows its simulation [4].

The effectiveness of simulation-based validation approaches is often not sufficient when critical systems are considered, where the domotic system must ensure some security -or safety- related properties, as is often the case when dealing with humans and their home environments. In this context, we advocate the use of Formal Verification techniques, that can give mathematical evidence that the desired system properties are satisfied by any possible system evolution.

This paper proposes a design time methodology for the Formal Verification of Intelligent Domotic Environments. The proposed approach is based on abstractly modeling the system as a set of concurrent UML 2.0 State Charts [5], that model the behavior of intelligent devices, the network connecting them, and the governing algorithms implemented in the home gateway. Such system is formally defined thanks to State Charts semantics [9], and can be formally verified by checking logical properties expressed in Temporal Logics [6, 7] by suitable model checking tools [11].

Formal Verification approaches are able to prove some properties of the provided models (in our case, state charts), but there is always the issue of ensuring the consistency of the verified model with the actual implemented system. This issue is solved in our approach thanks to the combination of:

1. ensuring the correctness of the state charts model of the devices and their communications using an automatic translation from the same system configuration description file that will be used at run time by the home gateway; this is based on the DogOnt [3] ontological description.
2. empowering the home gateway to interpret state charts at run time, so that the very same state charts for the governing algorithms that were used in formal verification will be used in the running system.

Verification techniques presented by J.C.Augusto at. el. [1] are based on Automata modeling of different devices IDE system, they also apply temporal properties for the verification of IDE system.

The remainder of this paper is organized as follows: basic description of the important components of the methodology is given in Section 2, and the main methodology is presented in Section 3 with a description of the adopted formalisms and tools. A case study showing the application of the methodology is presented in Section 4 and the associated verification results are discussed in Section 5. Concluding remarks and future work are finally discussed in Section 6.

2 Building Blocks

In this section, a brief introduction of DogOnt, Dog, State Charts and Temporal Logic is given, on which our verification methodology is based.

DogOnt [3] provides formal modeling and suitable reasoning for home environments through semantic web technologies. DogOnt is an ontological representation of the information of house layout and devices (which can be in real world IDEs) with their location, states and functionalities through semantic relations. The main focus is on the functionalities and states of the controllable devices in the IDEs.

Dog (Domotic OSGi Gateway) [2] is a domotic gateway for managing different domotic network components and their inter/intra communication and computational capabilities in a technology independent manner. Dog adopts the standard OSGi (Open Source Gateway initiative) [13] framework and it has capabilities for supporting the knowledge representation of semantic approaches and technologies; it can host intelligent applications.

State Charts are the Unified Modeling Language (UML) artifact, which is used for the graphical modeling of object-oriented softwares [5]. These are used for representing the dynamic aspect of the system, since the behavior of reactive systems can be modeled with the help of these State Charts.

DogSim [4] is an API for the automatic generation of State Charts. It takes the information of devices and their interconnection (event-messaging) from DogOnt, and with the help of Template Library Files (of devices) it translates this information into State Chart XML (SCXML) [14] files. These SCXML files are the state charts of devices and ready to be used for simulation or verification.

Temporal logic is a formalism widely used in Formal verification. It is a system of rules for reasoning with the different propositional quantifiers in terms of time. In the presented methodology, we use UCTL [12, 7], a UML-oriented branching-time temporal logic, which has the combined power of ACTL (Action Based Branching Time Logic) and CTL (State Based Branching time logic). UCTL uses the box operator ("necessarily," represented here as $[\![\,]\!]$) and the diamond one ("possible," $\ll\gg$) operator from Hennessy-Milner Logic and uses all temporal operators from CTL/ACTL (like Until, Next, Future, Globally, All, Exists).

3 Proposed Verification Methodology

As already stated, the goal of the proposed methodology is to verify functional properties of a given IDE, especially when it contains one or more control algorithms, which cause the overall system to exhibit complex behaviors. Our proposed approach, graphically summarized in Figure 1, is based on the following main assumptions:

1. the IDE is modeled according to the DogOnt ontology, that describes the Domotic plant composed of all the devices, their states, events and actions;
2. the control algorithms governing the IDE intelligence are expressed in the form of State Charts;

3. the residential gateway is able to process DogOnt system models and to interpret at runtime the State Charts describing the algorithms. In this paper, we adopt the Dog gateway [2] for this purpose;
4. system specification is given in the form of Temporal Logic properties and a suitable model checker is available for State Chart models.

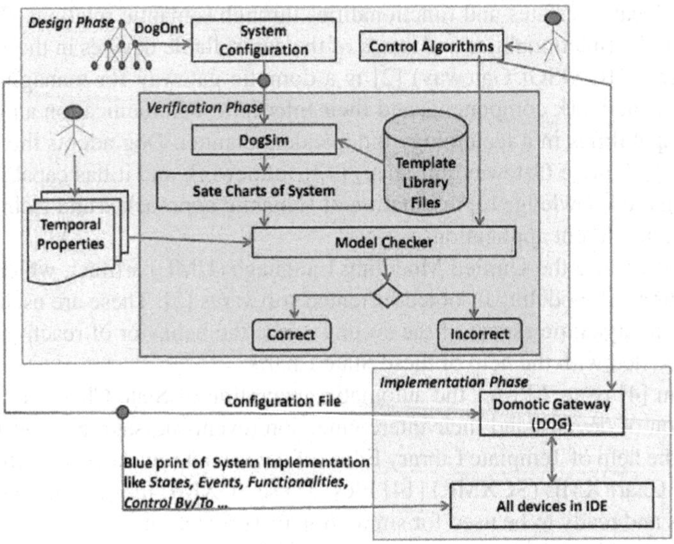

Fig. 1 Proposed Methodology

To apply model checking, we need a State Chart model of the domotic plant, corresponding to the DogOnt model. This can be obtained thanks to the DogSim compiler, that relies on a library of elementary State Chart templates describing each device's behavior. Such library templates are instantiated according to the devices listed in the DogOnt model and they are interconnected through 'connector' State Charts that model their communications and message exchanges; more details are available in [4].

Except for very trivial systems, the designer needs to implement one or more control algorithm(s) for embedding the necessary intelligence. We require such algorithms to be described in the form of State Charts. In general, deep analytical skills are required for the correct development of such control algorithm(s), and consequently for the correctness of the overall intelligence of the domotic environment.

The constraints and the behavior of the system, which is required to be checked, must be described in Temporal Logic. It is suggested [10, 11, 7, 12] that branching time Temporal Logic is best suited for the verification of State Charts, because it deals with multiple time lines. This manual property specification step requires a very skilled designer with the understanding of ACTL/CTL and Hennessy-Milner Logic. This step may require a significant amount of specification work, but is essential in the verification of any critical system.

Model checking can now be readily applied by verifying the temporal properties against the model consisting of the control algorithms coupled with the (automatically generated) State Charts of the domotic plant. The results of model checking will highlight the design errors, in the control algorithms or in the domotic plant, which should be located and corrected. The applicability of model checking to large systems needs to take into account the state explosion problem, and to address it with proper partitioning and abstraction techniques.

When all properties are successfully verified, we may safely move to the implementation step. The domotic plant, containing all the devices (that are assumed to be fault free), is connected with Dog, that allows the verified control algorithms to query and manage the actual devices. In fact, the control algorithms run by Dog are exactly the same State Charts used in the verification phase. Also, the same DogOnt model, that was used for generating the domotic plant model, is also used for Dog startup configuration. Therefore, the same information that was verified in the verification phase is used in the implementation phase, which guarantees that the system will work properly in all the verified scenarios.

4 Case Study

A simple but significant example of security and safety critical IDEs is a Bank Door Security Booth (BDSB). The BDSB system, represented in Figure 2, has two doors, for avoiding the harmful (direct) access of the user to the bank. One door is outside the bank and known as external door, whereas the other door is inside the bank and known as inner door. These doors have individual door-actuators for providing the force for opening and closing the doors. A user can access the door by pressing touch-sensors (TSs). These TSs are placed very near to doors and must be accessible on the each side of door.

The Door Lock Control (DLC) is the intelligent component of the BDSB system, it resides in the gateway (Dog in our case) and controls the evolution of the system intelligently. It checks the states of the devices and the events that devices generate; by considering the constraints, it decides what it has to do for obtaining correctness and security of the system. The DLC must ensure different constraints on the functionality of the overall system, some examples are: both doors can't be opened at same time; the request from any TS will not be processed again until it is acknowledged back after completing the task; a specific door will remain open for a definite time (after opening and before closing), so that, the people can cross the door, etc.

Let us consider a simple scenario, a user wanting to enter the bank: he/she can press T1 (the Touch Sensor on the outer side of the external door) and the request goes to the DLC. The DLC considers the current configuration of the system and (if all constraints are satisfied) it commands the external door actuator to open the door. Smoothly the external door will be opened until the door sensor provides the information to the external door actuator that the door is completely open. Now, the user can cross the door and will reach in the isolated space. After a given time, the DLC again commands the external door actuator for closing the door, and the door will be closed smoothly. For entering the bank he/she has to press T3, and the same door opening and closing process will be performed on the internal door and the user will be inside the bank. By pressing T4, the user can start the sequence for exiting the bank.

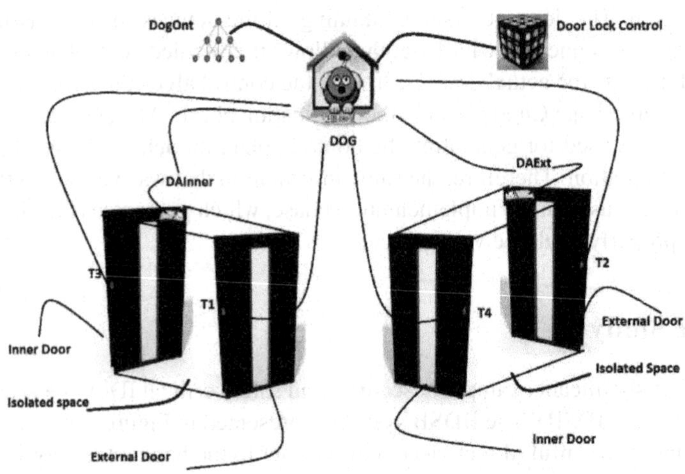

Fig. 2 Band Door Security Booth system with Dog

For the verification of these requirements, we select UMC (v 3.6) [10, 12] as a model checker because UMC supports state charts for describing the system, and for properties verification it can support UCTL.

5 Verification Results

After obtaining the State Charts of devices and control algorithm(s), it is required to translate these state charts into the input language of the chosen model checker: in our case, in the format of the UMC model checker. The designed model of the BDSB has 16,424 states. From the system specification, we derive different properties related to Security, Safety and Liveness on this model. For their verification, we evaluated nearly 50 different UCTL properties with the help of UMC framework. By considering the space limitation we here report just a few properties from the verified property set.

Property: *Against every posted request the specific TS must receive an acknowledgment*. The formula:

1. $AG[\![openRequest(T1)]\!]\ AF_{\{tsDone(T1)\}}\top$

is proven True by the model checker, which shows that if any sensor sends a door open request, sooner or later, it must receive an acknowledgment that the command has been executed.

Property: *TS will be available at anytime*. The formulas:

1. $AG[\![openRequest(T1)]\!]\top$
2. $AG[\![openRequest(T1)]\!]\ A\left[\top\ _{\{\neg openRequest(T1)\}}U_{\{tsDone(T1)\}}\ \top\right]$

are proven True by the model checker, which shows that TSs will be available at any time (*openRequest*), and a specific TS will not send another open request until it is acknowledged back after completing the task.

Property: *Interruption from any TS cannot break/change the execution of the current task*. The formulas:

1. $AG[\![openRequest(T1)]\!]\ AF[\![openRequest(T2)]\!]\ A\left[\top_{\{\neg daDoorOpen(DAExt)\}}U_{\{tsDone(T1)\}}\top\right]$
2. $AG[\![openRequest(T1)]\!]\ AF[\![openRequest(T3)]\!]\ A\left[\top_{\{\neg daDoorOpen(DAInner)}U_{\{tsDone(T1)\}}\top\right]$
3. $AG[\![openRequest(T1)]\!]\ AF[\![openRequest(T4)]\!]\ A\left[\top_{\{\neg daDoorOpen(DAInner)\}}U_{\{tsDone(T1)\}}\top\right]$

are proven True by the model checker. The set of three formulas satisfies the proper execution of T1, which means that when open-request of any door is in process and meanwhile another open request arrives from any other TS, the other request will not be executed until the first request completes its task.

Property: *Direct Access to the Bank is not possible*. The formulas:

1. $AG[\![daDoorOpen(DAExt)]\!]\ A\left[\top_{\{\neg daDoorOpen(DAInner)\}}U_{\{extDoorClosed()\}}\top\right]$
2. $AG[\![daDoorOpen(DAInner)]\!]\ A\left[\top_{\{\neg daDoorOpen(DAExt)\}}U_{\{innerDoorClosed()\}}\top\right]$

are proven True by the model checker, which shows that when one open-request is in process, if an open-request for the other door arrives, it can not be executed until the other door is closed, and therefore the doors cannot be open at the same time.

6 Conclusion and Future Work

Verification of the correct behavior of different heterogeneous devices integrated with control algorithms in Intelligent Domotic Environments improves safety, security and prevents critical threats. The presented methodology ensures the correct behavior of these IDEs with the use of Formal Model Checking techniques. In this paper, a small but not so simple case study is considered, and its correctness is proved against its temporal logic specification. The approach may also be applied to larger systems, even if scalability must be carefully ensured by partitioning and abstraction techniques usually adopted in model checking approaches.

In our future work, we plan to work on wider applicability and scalability of the approach, and on improving its automation. The problem of writing specifications will also be addressed, possibly by deriving temporal properties from ontology descriptions of the system.

Acknowledgements. This work is partially supported by the Higher Education Commission (HEC), Pakistan under UESTP-Italy/UET project. The authors thank Franco Mazzanti for his guidance and support.

References

1. Augusto, J.C., Mccullagh, P.: Ambient Intelligence: Concepts and Applications. Computer Science and Information Systems 4(1), 1–27 (2007)
2. Bonino, D., Castellina, E., Corno, F.: The DOG gateway: Enabling Ontology-based Intelligent Domotic Environments. IEEE Transactions on Consumer Electronics 54(4), 1656–1664 (2008), doi:10.1109/TCE.2008.4711217
3. Bonino, D., Corno, F.: DogOnt - ontology modeling for intelligent domotic environments. In: Sheth, A.P., Staab, S., Dean, M., Paolucci, M., Maynard, D., Finin, T., Thirunarayan, K. (eds.) ISWC 2008. LNCS, vol. 5318, pp. 790–803. Springer, Heidelberg (2008)
4. Bonino, D., Corno, F.: DogSim: A State Chart Simulator for Domotic Environments. In: 8th IEEE International Conference on Pervasive Computing and Communications Workshops (PERCOM Workshops), pp. 208–213 (2010), doi:10.1109/PERCOMW.2010.5470666
5. Booch, G., Rumbaugh, J., Jacobson, I.: Unified Modeling Language User Guide. The Addison Wesley, Reading (1998) ISBN 0-201-57168-4
6. Clarke, E.M., Emerson, E.A., Sistla, A.P.: Automatic Verification of Finite-State Concurrent Systems Using Temporal Logic Specifications. ACM Transactions on Programming Languages and Systems 8(2), 244–263 (1986)
7. De Nicola, R.: Three Logics for Branching Bisimulation. Journal of the Association for Computing Machinery 42(2), 458–487 (1995)
8. Ducatel, K., Bogdanowicz, M., Scapolo, F., Leijten, J., Burgelman, J.C.: Scenarios for Ambient Intelligence in 2010. Tech. rep., ISTAG: IST Advisory Group (2001)
9. Gnesi, S., Latella, D., Massink, M.: Modular semantics for a UML statechart diagrams kernel and its extension to multicharts and branching time model-checking. Journal of Logic and Algebraic Programming 51(1), 43–75 (2002)
10. Gnesi, S., Mazzanti, F.: On the fly model checking UML State Machines. In: ACIS International Conference on Software Engineering Research, Management and Applications, pp. 331–3382 (2004)
11. Gnesi, S., Mazzanti, F.: A Model Checking Verification Environments for UML Statecharts. In: Proceedings of the XLIII Congresso Annuale AICA (2005)
12. Mazzanti, F.: UMC 3.3 User Guide, ISTI Technical Report 2006-TR-33. ISTI-NNR Pisa-Italy (2006)
13. OSGi Service Platform release 4. Tech. rep., The OSGi alliance (2007)
14. State chart XML (SCXML): State Machine Notation for Control Abstraction. Tech. rep., W3C (2010), `http://www.w3.org/TR/scxml/`

Wheelchair-Based System Adapted to Disabled People with Very Low Mobility

Albano Carrera, Alonso A. Alonso, Ramón de la Rosa Steinz,
María I. Jiménez Gómez, and Lara del Val

Abstract. This paper presents a guidance system for wheelchairs, applied to severe motor disabilities, compatible with the commercial control systems. This allows adapting the interfaces developed for electric wheelchairs in common use, achieving a low cost of implementation. Several different interfaces adapted to fit the residual capabilities of various types of physical disabilities have been developed. The necessary electronic equipment has been built, the microcontroller's software has been programmed and the whole system has been tested to verify its operation. This equipment will replace conventional joystick function in a wheelchair. The interfaces are assembled on a commercial wheelchair and the performance of the whole system has been tested with some of the interfaces in control subjects. The tests were performed using a suitable protocol for this type of application. The ability of users to perform a predefined task and their ability to learn was measured, according to the protocol. Promising results were obtained in the execution of the test.

1 Introduction

Currently, there is a large number of people with disabilities that imply severe reduction of mobility, like tetraplegia, brain stroke or vascular brain damage. These people usually have impairments which prevent them from performing their normal daily activities. Thus the environment must be adapted to this type of disability to provide some degree of independence.

Albano Carrera · Alonso A. Alonso · Ramón de la Rosa Steinz
Laboratory of Electronics and Bioengineering, College of Telecommunications,
University of Valladolid, Paseo Belén, 15. 47011, Valladolid, Spain
e-mail: albano.carrera@uva.es, alonso@tel.uva.es

María I. Jiménez Gómez · Lara del Val
Array Processing Group, College of Telecommunications,
University of Valladolid, Paseo Belén, 15. 47011 Valladolid, Spain

P. Novais et al. (Eds.): Ambient Intelligence - Software and Applications, AISC 92, pp. 17–24.
springerlink.com

The interest of this field of work is established by the appearance of numerous different works and systems designed to make life easier for the user. In order for disabled people to function, one of the most studied tasks has been facilitating the control of personal computers [1] and [2].

As much as the work is focused on creating interfaces to control wheelchairs, there are several works that review the different existing systems and their conclusion is that improvements should be made on the interfaces of control and monitoring technologies for automatic navigation [3]. Other studies show similar systems to control wheelchairs. The solutions proposed in the literature can be grouped into the following strategies:

- Motion-detection camera systems, operated by the head or the eye, [4] and [5].
- Interfaces based on EMG and EOG records of muscle-generated signals for voluntary actions, [6], [7] and [8].
- EEG-based interfaces: Brain Computer Interfaces (BCI), [7], [8] and [9].
- Interfaces based on head position using inertial sensors, [10] and [8].
- Autonomous navigation aids for wheelchair users, [11], [12] and [13].
- Systems based on control by sniffing, [14].

The system presented in this paper is based on sensors capable of detecting winks performed voluntarily by the user. The hardware needed for the treatment of signals received is simple, robust, inexpensive, compatible with commercial wheelchairs and implemented by microcontrollers. The main advantages of this system are listed below:

- Simplicity of the hardware,
- Light software implemented on microcontrollers,
- More reliable than other systems,
- Low cost and easily adapted to different commercial wheelchair models,
- Minimal interference with patient activities. Necessary conventional biological flashes are not detected.

This paper has been divided into different sections: objectives, system structure, test and results and conclusions. These sections are explained below.

2 Objectives

The objectives of this study are to build interfaces adapted to control an electric wheelchair; these interfaces allow disabled people to use them. The control of the wheelchair is exercised through eye blinks and other movements made by patients with severe motor disabilities and that use voluntary residual capabilities. It seeks to create a system to replace or supplement efficiently the joystick incorporated in commercial chairs. The system allows progressive management of the speed and various maneuvers that are needed to move the wheelchair.

The next objective is to test the system performance. In principle the system has been tested using a simple user interface that consists of a keypad with four buttons that can be used by people with limited mobility.

The following tests are performed by an interface with a pair of glasses as support in order to detect small movements of the edge of the eye produced by voluntary winks. This system is particularly suitable for severe motor disabilities, as it takes advantage of the movement in the face and eyes only.

The last objective is to test the system with control subjects, before running it on actual patients, following a predefined custom protocol.

3 Structure of the System

The system consists of three distinct parts: the adapted interface, the processing system and the wheelchair (Fig.1). This section will explain each of these elements and software features included in the processing system.

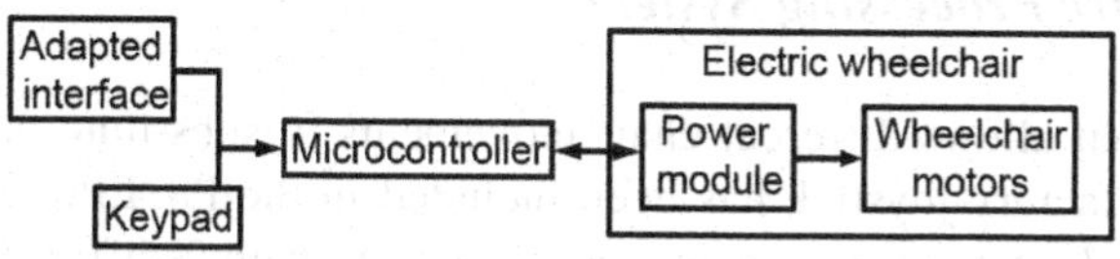

Fig. 1 Block diagram of the implemented system

3.1 Adapted Interface

Different types of adapted interfaces have been designed and built. Most of them required an additional support consisting of a pair of glasses and mechanical items for attachment and adjustment for different sensors. Then, each of the sensors used are described:

- Optical sensors based on a popular integrated circuit CNY70. The sensor is placed on the arm of the glasses and a black/white sticker on the side of the face near the edge of the orbicularis oculi, [15]. Following this philosophy, two types of systems for the acquisition of signals have been built: a system based on PIC16F84A microcontroller from Microchip Inc and a system based on the popular Arduino hardware platform.
- Optical sensors based on the system used in conventional optical mice, placed on the arm of the glasses. The advantage of this new sensor is that it does not need any special attachments, simply put it in the right place and it functions properly. The system is controlled by an Arduino.
- Vibration sensors based on the piezoelectric effect. These sensors have been built in conjunction with a signal conditioner which adapts the signal to a PIC microcontroller.

- Electromyogram electrodes for acquisition of contraction signals of the orbicularis muscle. These interfaces need surface electrodes.
- Keypad. This device allows an intermediate interface harder to use by a disabled person to control the wheelchair. The keypad provides at its output the same commands that have been adopted as standard for the other interfaces.

3.2 Hardware Processing System

A low-power, low-cost, P87LPC769 microcontroller from NXP Semiconductors Company was used as processing system. This device has four analog inputs, two outputs with digital/analog converter, a UART interface and the possibility of configurations for two external interruptions and allows an integrated 5-volt supply. The system is configured with an external clock of 20 MHz and all peripherals and connectors to ensure proper operation.

3.3 Software Processing System

A program that allows different chair movements to substitute the role of the incorporated factory joystick has been included in the P87LPC769 microcontroller. Thus, 7 types of movements are supported with 7 different orders: advance and move back, turn to the right/left, turn slightly forwards to either side, and stop. 5-speeds have been incorporated to allow better adaptation and management.

To achieve this behaviour, identical signals to those generated in the joystick input interface have been synthesised. By controlling the phase of the signals produced, the system can govern the sense of rotation of the motor and the amplitude controls the movement speed. Fig. 2 (left) shows the different signals used by the system. The central signal of each movement corresponds to a sync signal read by the microcontroller from the control system of the wheelchair. The other two are used to change the motion of the motor.

3.3 Electric Wheelchair

Basically, a commercial wheelchair consists of a power module responsible for motor control, a joystick or control device, which is used as an interface for transmitting the orders of the patient, and motors to put the wheelchair in motion, Fig. 2 (right).

In this case a SHARK wheelchair control system from Dynamic Controls Company is used. This system consists of a power module and a joystick. The joystick is *DK-REMA* [16], while the power module is *DK-PMA* [17]. The latter supplies power to the motors of the wheelchair and locks and unlocks the brakes.

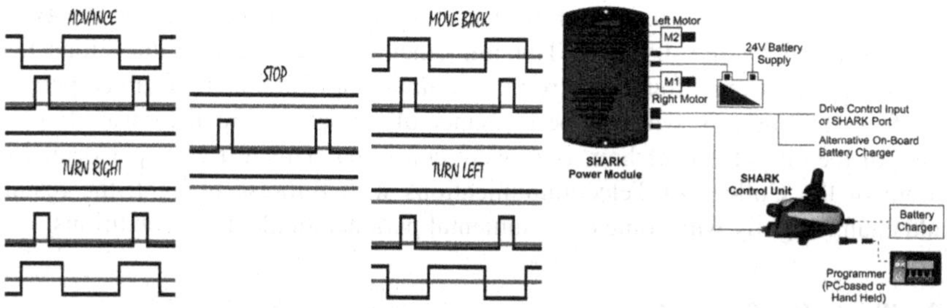

Fig. 2 Left: Control signals synthesized by the microcontroller. Right: General wiring diagram of the commercial wheelchair.

4 Tests and Results

After verification of hardware, overall system performance was tested with all the interfaces described in section 3.1. Satisfactory results were achieved as far as the adaptation between the developed hardware and the wheelchair control commercial system, the system is fully functional and the results were as expected. A specific protocol for conducting these tests has been developed. These trials have proven that, after a short training process, a user can perform optimal control of the wheelchair.

4.1 Protocol

A testing protocol to verify the goodness of the developed system and the users' ability and their improvement after a period of learning was defined. This protocol is based on completing a circuit in a space inside a room and measure the time spent exercising, Fig. 3, in three turns. In the first trial, the user makes the circuit with no previous experience in using the interface. Then the user can have free training on his/her own for five minutes. After the workout, the user repeated the circuit twice in a row, and the improvements in the time taken to do so were observed.

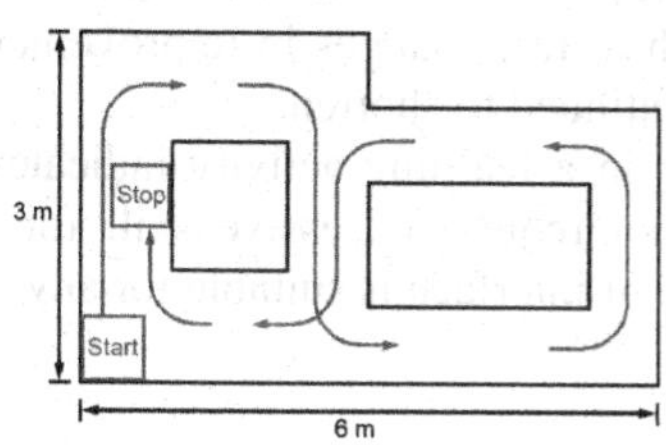

Fig. 3 Circuit designed to perform tests on users. The dimensions of the circuit are not stated for clarity, but they are similar to a conventional room.

The circuit designed uses real obstacles to achieve maximum similarity to situations faced by the disabled. The implemented circuit used dimensions that mirror those that can be found in a room of a non-adapted home to disabled people. Therefore, it allows assessing the performance of the task in a realistic way. For the design of the circuit, wheelchair users were consulted. This test were performed in a room of the College of Telecommunications with fluorescent light, the results could change lightly with some environmental parameters like light conditions.

4.2 *Tests Performed*

The system was tested with a control population of 11 individuals of different ages and sex. None of them had previously used the system. The ages ranged between 22 and 48 years old, photo in Fig. 4.

The measured times in the first round, when users have no previous training, and the measured times in the two subsequent repetitions, when subjects had trained for 5 minutes, are shown in the figure below, plot in Fig. 4.

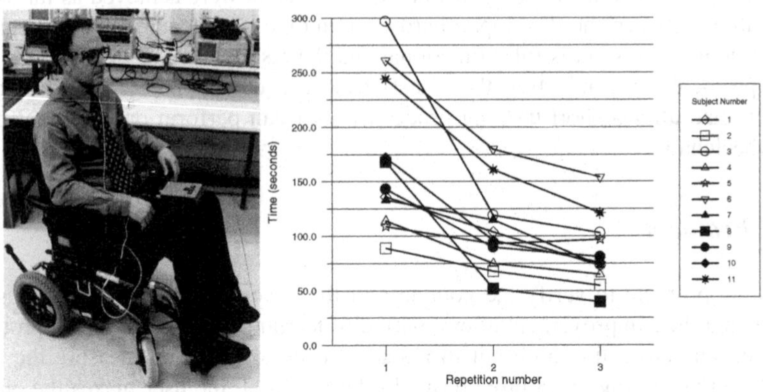

Fig. 4 Left: Photo of a patient during the tests. Right: Plot that represents the time, in seconds, taken by each subject to complete the proposed circuit in each of the three repetitions.

Similar behaviour was observed with respect to learning in all users. We noted a great improvement between the first and the second round. On the other hand, between the second and third test changes in improvements were less noticeable, but the execution times continued to shorten.

This type of behaviour in a learning activity indicates that the task is understood quickly and does not require excessive skill for its proper performance, which shows that this type of interface is suitable for any user.

5 Conclusions

The system presented in this paper can use different types of interfaces adapted to guide a wheelchair for people with severe motor impairment. Using this system, people who have very limited mobility but are able to make small movements with the muscles of the face can manipulate the electric wheelchair independently. This system improves the possibilities presented by the group in previous works [18-20] where they performed the guidance of a robot. In this case, a commercial wheelchair was used, which increases the possibilities of the system.

The different interfaces implemented for the operation of the chair proved to be suitable for use by people with reduced mobility. The data collected after the tests with control subjects employing the keypad interface and the protocol designed demonstrate that communication between our system and the trading system works robustly and effectively. In addition, it was found that all interfaces based on the detection of winks functioned correctly, allowing the user great flexibility of movement in their environment. Besides, the operating principle of this system allows users to perform other tasks simultaneously with minimal interference in the guidance. If the system is compared with other types of interfaces, such as BCI, it can be concluded that the system is faster, more reliable, requires less computational load and is more convenient to use.

The system is operational and ready for testing on patients with physical disabilities or brain sickness. Similarly, the interface and built-in system can also be applied to different home automation devices that offer different functions within the usual environment of the disabled.

Acknowledgments. This work has been funded by the Research Excellence Group GR72 of the Junta de Castilla y León.

References

1. Gareth Evans, D., Drew, R., Blenkon, P.: Controlling mouse pointer position using an infrared head-operated joystick. IEEE Transactions on Rehabilitation Engineering 8(1), 107–116 (2000)
2. Betke, M., Gips, P., Fleming, P.: The Camera Mouse: Visual Tracking of Body Features to Provide Computer Access for People With Severe Disabilities. IEEE Trans. Neural Syst. Rehabil. Eng. 10(1), 1–10 (2002)
3. Fehr, L., Langbein, W.E., Skaar, S.B.: Adequacy of power wheelchair control interfaces for persons with severe disabilities: A clinical survey. IEEE Journal of Rehabilitation Research and development 37(3) (2000)
4. Úbeda, A., Azorín, J.M., Iáñez, E., Sabater, J.M.: Interfaz de seguimiento ocular basado en visión artificial para personas con discapacidad. In: III International Congress on Domotics, Robotics and Remote-Assistance for All - DRT4all2009, Barcelona, Spain, pp. 25–33 (2009)
5. Perez, E., Soria, C., Mut, V., Nasisi, O., Freire Bastos, T.: Interfaz basada en visión aplicada a una silla de ruedas robótica. In: III International Congress on Domotics, Robotics and Remote-Assistance for All, Barcelona, Spain, pp. 33–45 (2009)
6. Barea, R., Boquete, L., Mazo, M., López, E.: System for assisted mobility using eye movements based on electrooculography. IEEE Trans. Neural Syst. Rehabil. Eng. 10(4), 209–218 (2002)

7. Ferreira, A., Cardoso Celeste, W., Freire bastos-Filho, T., Sarcinelli-Filho, M., Auat Cheein, F., Carelli, R.: Development of interfaces for impaired people based on EMG and EEG. In: II International Congress on Domotics, Robotics and Remote-Assistance for All, Madrid, Spain, pp. 187–196 (2007)
8. Freire Bastos, T., Ferreira, A., Cardoso Celeste, W., Cruz Calieri, D., Sarcinelli Filho, M., de la Cruz, C.: Silla de ruedas robótica multiaccionada inteligente con capacidad de comunicación. In: III International Congress on Domotics, Robotics and Remote-Assistance for All, Barcelona, Spain, pp. 45–52 (2009)
9. Iturrate, I., Escolano, C., Antelis, J., Minguez, J.: Dispositivos robóticos de rehabilitación basados en Interfaces cerebro-ordenador: silla de ruedas y robot para teleoperación. In: III International Congress on Domotics, Robotics and Remote-Assistance for All, Barcelona, Spain, pp. 124–134 (2009)
10. Azkoitia, J.M., Eizmendi, G., Manterota, I., Zabaleta, H., Pérez, M.: Non-Invasive, Wireless and Universal Interface for the Control of Peripheral Devices by Means of Head Movements. In: II International Congress on Domotics, Robotics and Remote-Assistance for All, Madrid, Spain, pp. 211–220 (2007)
11. Levine, S.P., Bell, D.A., Jaros, L.A., Simpson, R.C., Koren, Y., Borenstein, J.: The NavChair Assistive Wheelchair Navigation System. IEEE Transactions on Rehabilitation Engineering 7(4), 443–451 (1999)
12. Minguez, J., Montesano, L., Díaz, M., Canalis, C.: Intelligent robotic mobility system for cognitive disabled children. In: II International Congress on Domotics, Robotics and Remote-Assistance for All, Madrid, Spain, pp. 235–242 (2007)
13. Angulo, C., Minguez, J., Díaz, M., Cabestany, J.: Ongoing Research on Adaptive Smart Assistive Systems for Disabled People in Autonomous Movement. In: II International Congress on Domotics, Robotics and Remote-Assistance for All, Madrid, Spain, pp. 297–304 (2007)
14. Plotkin, A., DSela, L., Weissbrod, A., Kahana, R., Haviv, L., Yeshurun, Y., Soroker, N., Sobel, N.: Sniffing enables communication and environmental control for the severely disabled. Proceedings of the National Academy of Sciences of the United States of America 107(32), 14413–14418 (2010)
15. Campos, M.A.: Mejoras para sistemas de interfaz hombre-máquina aplicados al guiado de un microbot y a la comunicación aumentativa. Master thesis. College of Telecommunications. University of Valladolid., Valladolid (2010)
16. Installation Manual SHARK DK-REMA Remotes, Dynamic Controls (2006)
17. Installation Manual DK-PMA SHARK Power Module, Dynamic Controls (2004)
18. Alonso, A., De la Rosa, R., Hornero, R., Abásolo, D.: Estudio de la viabilidad de un sistema de sillas de ruedas autoguiadas (SRAu) para discapacitados. In: XXII Congreso anual de la sociedad española de ingeniería biomédica - CASEIB 2004, Santiago de Compostela, Spain (2004)
19. Alonso, A.A., de la Rosa, R., del Val, L., Jiménez, M.I., Franco, S.: A robot controlled by blinking for ambient assisted living. In: Omatu, S., Rocha, M.P., Bravo, J., Fernández, F., Corchado, E., Bustillo, A., Corchado, J.M. (eds.) IWANN 2009. LNCS, vol. 5518, pp. 839–842. Springer, Heidelberg (2009)
20. Alonso, A., Jiménez, M.I., Del Val, L., De la Rosa, R., Izquierdo, A.: Design and implementation of a control and guided system for mobile device by means of interfaces adapted to severe physical disability. In: International Symposium on Ambient Intelligence, Guimarães, Portugal, pp. 55–62 (2010)

Toward Seamless Environments for Dispute Prevention and Resolution

Davide Carneiro, Paulo Novais, and José Neves

Abstract. Given the evolution of the Information Technology society, it is now rather simple to acquire products or services in a foreign country. This practice may conduct to the event of conflicts whenever a consumer detects some fault or malfunction in services or products he/she had bought. A situation that may worsen if at the time of the uncovering of the defect, the shopper is already in a different geographical arena. There is thus the need to develop computational tools that may prevent these disputes from even happening. In this work it is proposed the development of seamless intelligent environments for dispute resolution that will surround the user, independently of his/her location. It is described the implementation of a prototype that may provide contextualized real-time information and legal support to consumers. The objective is to decrease the number of disputes due to a poor understanding in relation to the The Law and make justice more personalized and closer to people.

Keywords: Online Dispute Resolution, Mobile Online Dispute Resolution, The Law, Intelligent Environments.

1 Introduction

The technological developments in the last decades led to undeniable changes in our society, visible in barely all aspects of our daily lives. One of those aspects is the way in which we buy products. While in the past goods or services were bought in person in a relatively small geographical area, it is now possible to buy them from any part of the world. The most common way is to use online stores such as Amazon or eBay, which truly enabled worldwide commercial transactions. On the other hand, it is also possible to travel abroad and acquire those same

Davide Carneiro · Paulo Novais · José Neves
Department of Informatics
University of Minho, Braga, Portugal
e-mail: {dcarneiro,pjon,jneves}@di.uminho.pt

P. Novais et al. (Eds.): Ambient Intelligence - Software and Applications, AISC 92, pp. 25–32.
springerlink.com

goods or services in a foreign country. In any way, there is a normal risk on this kind of transactions, i.e., buyers are not always aware of their rights as consumers.

Indeed, problems arise when the product that was bought does not comply either with the seller marketing, lacks some feature, is of poor quality or is even defective. In such circumstances, the regulations that apply are the rules of the country in which the store is located, and the consumer may be faced with some problems, namely:

(i) Once a buyer acquires an item, generally he/she does it abroad, and normally is not aware of his/her rights as a consumer; and
(ii) When the store does not wish for to settle the affair, and the buyer decides to go into a litigation process.

On the other hand, such processes may have to be conducted in a unfamiliar environment, making it impractical for cut-rate products. If we acknowledge that the majority of online transactions deal with such products, we can estimate the amount of small-value disputes that arise every day and are not worked out because of unsuited legal processes.

Therefore, two main problems may be identified: (1) the buyer is usually unaware of his/her consumer rights; and (2) the buyer has no realistic way of solving an eventual conflict. Ultimately, this may lead to a decrease in the degree of satisfaction of the buyers, influencing in the negative the business-related transactions.

This trend comes with other challenges that face the legal field. Know-how, in general, has significantly increased the amount of disputes, rendering courts slow and ineffective, making access to justice more difficult and thus less fairly. One answer to this problem may came in the form of Alternative Dispute Resolution (ADR), a way of solving disputes out of courts, using alternative processes such as negotiation, mediation or arbitration [1, 2, 3]. Nevertheless, these processes still require the disputant parties to meet in a physical place in order to solve the dispute. The evolution towards Online Dispute Resolution (ODR) is therefore regarded as a natural way [4]. ODR describes a set of methods in which technology is used to implement already traditional forms of ADR, in online environments. In order for ODR to be more than simple negotiation or mediation over a "phone line", artifacts from Artificial Intelligence may have to be considered [5]. The objective is that ODR tools may be able to support disputant parties by actively proposing strategies and solutions, compiling useful information, ultimately making judicial processes more efficient.

However, in order to interact with such ODR tools, parties still need to have access to web-enabled computational platforms. Moreover, these tools are usually only used to solve a dispute. In this work, we look to this problem under a different perspective, namely at a seamless environment for dispute prevention and resolution. The main objective is to build up an environment that may, in principle, to prevent a dispute by providing key information at the time of the purchaser, about its possible negative results, taking into consideration the norms that apply in the present customer scene, i.e., when the buyer actually buys the item, he/she may be aware of the potential future costs in cases of product defect. The secondary objective of this work is to provide mobile access to a previously existing ODR platform, i.e., UMCourt Commerce [6]. UMCourt Commerce is an agent-based [7] ODR platform targeted at the Commercial Law field. Using this online platform, an unsatisfied buyer can fill in a complaint and the platform suggests an outcome based on the Portuguese Commercial Law. A detailed description of the platform is provided in [6], including a description of the process model, some use scenarios and the field of law covered, which includes Decree of Law (DL) 67/2003 (as published by DL 84/2008) (Portuguese laws) [8, 9].

In the long term, we want to complement UMCourt with Ambient Intelligence functionalities [12] that may provide even more circumstantial information for the dispute resolution process, including the parties' emotional state. These environments will thus be abstract in essence, in order to provide such information regardless the domain of the dispute. In that sense, this work is being developed in three different legal domains, namely The Labor Law, Property Division and, as mentioned before, The Commercial Law.

2 Ambient Intelligence for Dispute Resolution

Among the disadvantages of current ODR tools, one of its major drawbacks is their coolness, i.e., it is easier for parties to lie without the intimidating presence of a judge. Moreover, a very important part of the communication process is lost. Mehrabian [14], states that most of the meaning that we derive from a face-to-face conversation comes from other facets than the words spoken, namely the tone of voice, the loudness, the facial expressions or the body gestures. Indeed, it is our conviction that intelligent environments would allow, at all, that this knowledge is to be acquired and included in the dispute resolution process model, making it richer, fairest, trueness and more effective.

We are currently taking the first steps towards the integration of Ambient Intelligence techniques with ODR platforms, resulting in dispute resolution environments that may support these and other functionalities. At this point, this environment tends to support wireless networks and a mobile device on the client side, holding up on an application server. Following this approach, and considering the Commercial Legal Domain, we want buyers to use the tool with the objective of getting information about their rights and risks as consumers, prior to buying a product or service. In that sense, the prototype developed can be used in two different ways, i.e., it can be used as an interface for the pre-existing platform, or it can be accustomed as a protective information tool. The main objective of the mobile device is to gather as much background information as possible (e.g., current realm, regional settings) in order to make available helpful actions or work for customers.

2.1 The Architecture and an Overview of the Environment

The development of such architecture poses some multifaceted challenges. Indeed, the most palpable one is how to interconnect (remote) software agents with the ones in the environment. In order to address this problem, the architecture adopted was based on two open technologies: OSGi and Jade. The development process followed in order to combine these two technologies to build an intelligent environment is described in [13]. OSGi is a modular approach to software development that relies on the concept of bundle: a bundle is able to provide and use services. It is used to fulfill a main goal, that of creating a consistent layer that will allow any device to be connected with the software layer. This layer is twisted by assigning (at least) one bundle for each appliance (e.g., sensor, personal device). Each bundle is specialized in that appliance, hides its singularities and provides its functionalities in terms of an OSGi service. These functionalities may then be used by any other bundle, without these being specialized on how to use a particular one, i.e., they simply have to request a standard service.

Jade agents [11], on the other hand, are used to implement highlevel behaviors (e.g., rule-based reasoning, case-based reasoning, decision making). Concretely, given the legal domain being focused, agents contain knowledge about the Portuguese Commercial Law, in the form of a set of laws. These agents thus take the information about the case and, depending on the type of the request, evaluate it and return a solution. Agents are tendentiously on the server side.

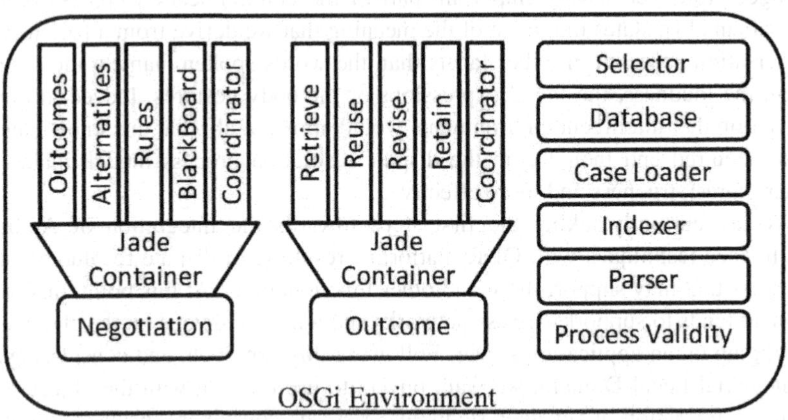

Fig. 1 A simplified view of the architecture. A more detailed description is given in [13].

On the client side, an application was developed to act as an interface between the user and the server. This function allows the user to login or register on the environment (Figure 2). The same login information is used either on the mobile or the web site, creating a seamless information environment. This means that the user may access his/her data (e.g., current case, previous cases, personal information) using a standard web browser, or he/she may access a lighter version of it and particularities on the dispute, under an ODR environment.

Table 1 The system information

Information	Description
Value (new)	The value of the product or service when it was bought
Value (used)	The current estimated commercial value of the product or service
Type of product	The nature of the product or service (e.g., real estate, mobile, used, new).
Date of purchase	The date under which the product or service was made available to the customer.
Date of event	The date under which the defect was detected by the customer.
Date of complaint	The date under which the buyer communicates the defect to the seller.
Delivery to Supplier	The date under which the product is delivered to the seller, for repair.
Temporary replacement	The date under which a similar product or service is made available to the user to be used as a substitution of the product or service being repaired.
Product return	The date under which the product or service was returned to the customer, after reparation.
Event description	A description of the type of defect detected in the product or service.
Seller description	A description of the seller.
Additional information	Additional information such as warranties or recipes.

The computational construction implements three main functionalities: (1) it allows the user to manage his/her personal information (e.g., contact information, address); (2) it permits the user to consult the solutions of previous cases, and (3) it lets the user to fill in a new case. When filling in a new case, the user can do it in two different ways, depending on whether he/she wants to simulate a given defect on a product or service he/she is considering buying, or there is an actual defect with a product that he/she has already bought. The table depicted below denotes the information that may be necessary for filling in a new case. Note that not all the information is mandatory, depending on the type of the request.

When a request is sent to the server, the agents build a new solution. Once again, depending on the type of the request, the solution is provided in two different ways. If the user is simply simulating an eventual malfunctioning (the simplest case), the solution is provided right away by the agents and shown on the mobile device. If, on the other hand, the user is looking at a solution in a case which may involve access to more multifaceted structured data, the solution is not provided in a hurry, once it must be validated by a human expert, as detailed in [6], i.e., the solution will be available later, after validation. Thus, the user can access it through the mobile device or through the web site.

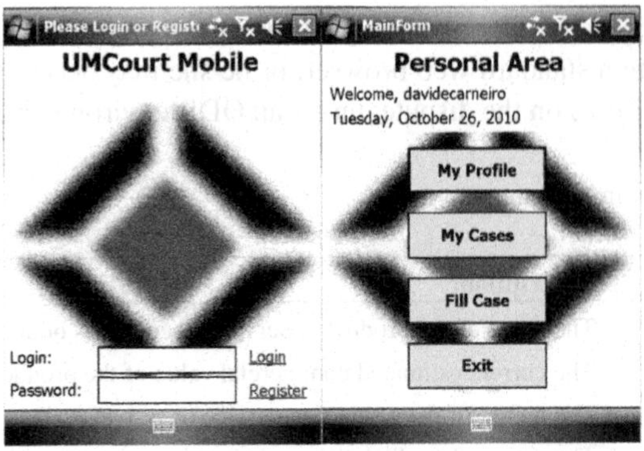

Fig. 2 The login interface and the application main menu

2.2 A Scenario

In order to highlight the main functionality of the application being developed so far, let us consider an example, as it is tinted on Figures 3 and 4; i.e., John, an English citizen, recently moved to Portugal for a few months and bought a PDA for domestic/private use. The contract that was celebrated is of buy/sell type. October 22nd, 2010, was the product delivery date. John finds a defect on October 26th, 2010, and wants to guess the possible outcome if he decides to deliver the PDA to repair and/or substitution on October 30th, 2010. As evidence, John uploads a warranty and a receipt, relative to the dates mentioned above. Concerning the defect, the buyer claims that the product does not fully comply with what was advertised (e.g., the seller said that the PDA has a GPS receiver, but it does not appear to be working). The supplier acts within the range of his/her professional skills and he/she is the producer of the artifact.

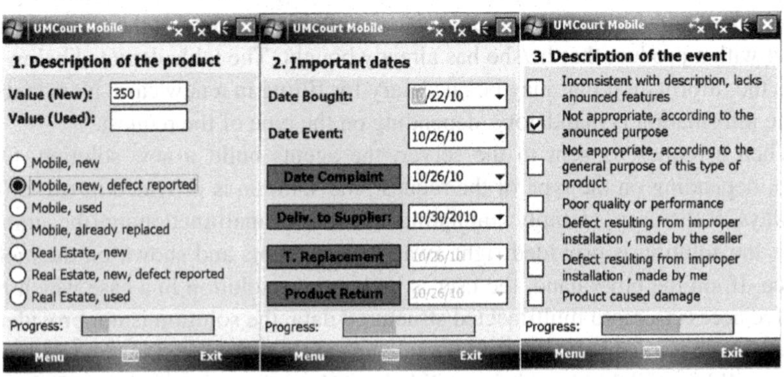

Fig. 3 The first three instances of the interface used for filling in a new case

When the customer demands an explanation, there is some extra data that must be taken into consideration, namely his/her name and/or e-mail address, obtained on or after the system, and his/her current location, acquired from the mobile operator. An XML file is then created that includes all the information so far attained which is stored in the user device (Figure 5). This, in turn, is received by the server, which analysis all the legal time-frames that apply in such a particular case (i.e., the software agents make sure that the legal time-frames are under the warranty time-frame (11 days), that the limit of two months counted up since the date of the defect detection has been respected (7 days), and that less than two years passed since the date of the complaint). As the product was delivered for repair and/or substitution and all the legal time-frames have been respected, John is informed that the supplier has two choices: either repair the product in 30 days (at most) at no cost or to proceed to the replacement of the product by an equivalent one (Figure 5). At this moment, John is aware of his rights regarding the current matter, according to the legal system of the country he is currently on.

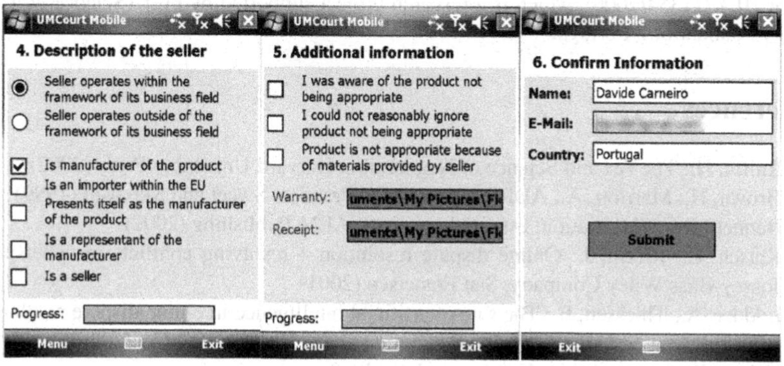

Fig. 4 The last three instances of the interface used for filling in a new case

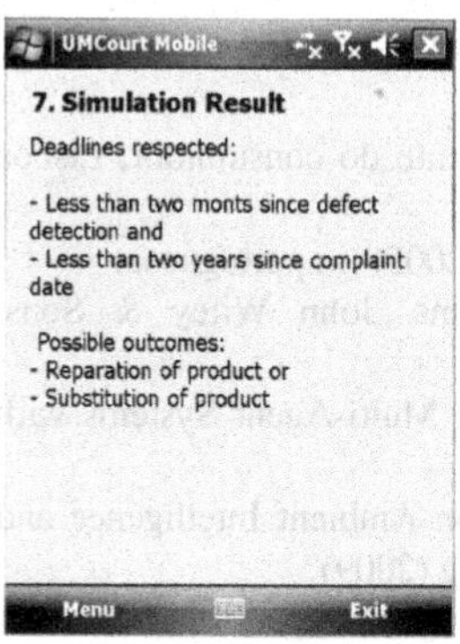


```
<?xml version="1.0"?>
<!--Information requested in 26-10-10 12:22:04-->
<case_data>
    <value_new>350</value_new>
    <product_type>2</product_type>
    <date_bought>22-10-10</date_bought>
    <date_event>26-10-10</date_event>
    <date_delivery>30-10-10</date_delivery>
    <event_description>2</event_description>
    <within_framework>true</within_framework>
    <seller_desc>1</seller_desc>
    <info>1</info>
    <warranty>\My Documents\As Minhas Imagens\~warranty.pdf</warranty>
    <receipt>\My Documents\As Minhas Imagens\~receipt.pdf</receipt>
    <user_name>Davide Carneiro</user_name>
    <user_mail>[illegible]</user_mail>
    <country>Portugal</country>
</case_data>
```


Fig. 5 A snapshot of an interface showing the result of the simulation and the XML file corresponding to the scenario considered

3 Conclusions

As it has been seen in previous work, in general, rule-based systems are ideal to be used in definite legal domains, such as The Commercial Law. Indeed, a rule-based ODR system was extended under this setting, with a mobile tool that allows the user to access it, independently of his/her location. Moreover, this tool adapts the solutions provided according to the local legal system. What is proposed is a shift from a paradigm in which the user has to explicitly interact with an ODR tool in a specific manner, to a paradigm in which the ODR tool is always available to the user, independently of his/her location. Moreover, such tool adapts its response according to the user location. Thus, we can now speak of virtual environments for dispute resolution, empowered by mobile devices that will significantly improve and personalize access to The Law.

Acknowledgments. The work described in this paper was developed under the TIARAC - *Telematics and Artificial Intelligence in Alternative Conflict Resolution* Project (PTDC/JUR/71354/2006), which is a research project supported by FCT (Science & Technology Foundation), Portugal.

References

1. Raiffa, H.: The Art and Science of Negotiation. Harvard University Press (2002)
2. Brown, H., Marriott, A.: ADR Principles and Practice. Sweet and Maxwell (1999)
3. Bennett, S.C.: Arbitration: essential concepts. ALM Publishing (2002)
4. Katsch, E., Rifkin, J.: Online dispute resolution – resolving conflicts in cyberspace. Jossey-Bass Wiley Company, San Francisco (2001)
5. Lodder, A., Thiessen, E.: The role of artificial intelligence in online dispute resolution. In: Workshop on Online Dispute Resolution at the International Conference on Artificial Intelligence and Law, Edinburgh, UK (2003)
6. Costa, N., Carneiro, D., Novais, P., Andrade, F.: An Agent-based Approach to Consumer's Law Dispute Resolution. In: Proceedings of the 12th International Conference on Enterprise Information Systems - ICEIS 2010 (2010)
7. Weiss, G.: Multiagent Systems: A Modern Approach to Distributed Artificial Intelligence. The MIT Press, Cambridge (2000)
8. Almeida, T.: Lei de defesa do consumidor anotada, Instituto do consumidor, Lisboa (2001) (in portuguese)
9. Almeida, C.F.: Direito do Consumo. Almedina. Coimbra (2005) (in portuguese)
10. Wooldrige, M.: An Introduction to Multiagent Systems. John Wiley & Sons, Chichester (2002)
11. Bellifemine, F.L., Caire, G., Greenwood, D.: Developing Multi-Agent Systems with JADE. Wiley, Chichester (2007)
12. Augusto, J.C., Nakashima, H., Aghajan, H.: Handbook on Ambient Intelligence and Smart Environments: a state of the art. Springer, Heidelberg (2009)
13. Carneiro, D., Novais, P., Costa, R., Neves, J.: Developing intelligent environments with oSGi and JADE. In: Bramer, M. (ed.) IFIP AI 2010. IFIP AICT, vol. 331, pp. 174–183. Springer, Heidelberg (2010)
14. Mehrabian, A.: Silent Messages – A Wealth of Information about Nonverbal Communication. Personality & Emotion Tests & Software, Los Angeles (2009)

Integrating Personalized Health Care and Information Access for Elder People

Diego Gachet Páez, Juan R. Ascanio, Ignacio Giráldez, and Margarita Rubio

Abstract. The concept of the information society is now a common one, as opposed to the industrial society that dominated the economy during the last century. It is assumed that all sectors should have access to information and reap its benefits. Elder people are, in this respect, a major challenge, due to their lack of interest in technological progress and their lack of knowledge regarding the potential benefits that information society technologies might have on their lives. The Naviga Project (An Open and Adaptable Platform for the Elder people and Persons with Disability to Access the Information Society) is an European effort whose main goal is to design and develop a technological platform allowing elder people and persons with disability to access the Internet and the Information Society. NAVIGA also allows the creation of services targeted to social networks, mind training and personalized health care.

Keywords: elderly, wellbeing, ambient assisted living.

1 Introduction

Today, developed countries have great difficulties with effective health services and quality of care in a context marked by the population's ageing. This trend, as seen in Figure 1, has dramatic effects on both public and private health systems, as well as on emergency medical services, mainly due to an increase in costs and a higher demand for more and improved benefits for users, as well as for increased person's mobility.

Diego Gachet Páez · Ignacio Giráldez · Margarita Rubio
Universidad Europea de Madrid, 28670 Villaviciosa de Odón, Spain
e-mail: {diego.gachet,ignacio.giraldez,margarita.rubio}@uem.es

Juan R. Ascanio
Encore Solutions C /Cronos 20, 28037 Madrid, Spain
e-mail: juan.ascanio@encore.es

P. Novais et al. (Eds.): Ambient Intelligence - Software and Applications, AISC 92, pp. 33–40.
springerlink.com

This demographic change will lead to significant and interrelated changes in the health care sector and technologies promoting independence for the elderly's. As representative data, approximately 64% of the European population is made up of 20 to 64 year olds, while the 65 and over group covers 17%. Thus, there are some 4 working employees to every pensioner. On the other hand, it is estimated that the 20 to 64 year old group will decrease to 55% and the over 65 will increase to 28% by the year 2050, making the proportion 1 to 2 instead of 1 to 4. Spending on pensions, health and long-term care is expected to increase by 4-8% of the GDP in the coming decades, with total expenditures tripling by 2050.

People live longer in developed countries as a result of better living and health conditions. For example, in North America only 4.5 % of population over 65 years old lives in nursing homes, a percentage that has decreased in recent years.

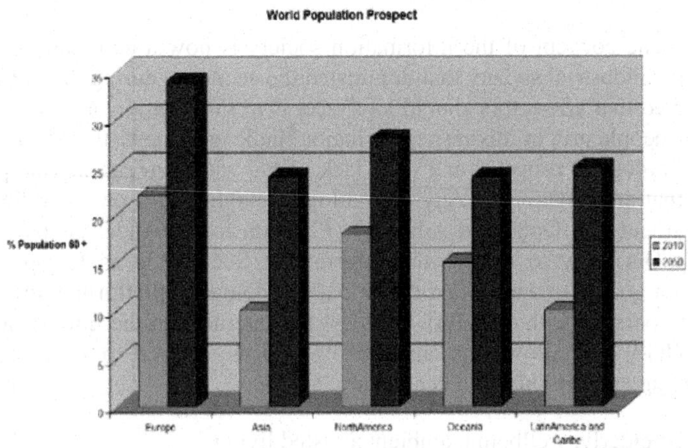

Fig. 1 Demographic change according to the foresight of the United Nations, http://esa.un.org/unpp (access: 06/12/2010)

The elder population is constantly prepared for to age better due to a decrease in disability, resulting in the elderly's being more active in their daily lives. Despite the improvement in conditions for coping with ageing and an increasingly active lifestyle, there are obvious changes that occur in behaviours and skills during the latter part of life.

These changes may include decreases in social relations and physical abilities, loss of memory, comprehensive and cognitive functions. Previous studies have shown that the ageing process is accompanied by a decrease in neuro-motor and cognitive functions. Compared to young people, the elderly's demonstrate poor performance on tests, including reaction times, motor coordination, short-term and complex or abstract conceptualization. In general, these changes result in a decline in the quality of life for the elderly's.

Another important impact that can be seen particularly in persons living in nursing homes is boredom. Participation in social activities does not necessarily improve this feeling and sometimes creates negative attitudes in participants, although activities based on individual preferences can have positive effects and help to overcome boredom, increasing the quality of life for elderly's. It is a challenge finding innovative activities that involve the elderly's and encourage them to keep practicing with the activity. An adequate understanding of the disuse of motor and cognitive functions can help to prevent the decline in these skills and participation in activities based on individual preferences can reduce boredom. There is a real need for activities that address these two concepts, and these activities may be none other than for example mental exercises and social networks specially designed for elderly's.

The current trend is to improve the quality of life of older people not only extend the lifetime, the "gerontechnology" [1] is a very active discipline focused on improving the lives of elderly's, considered as a special group of users whose particular skills and needs in social and cognitive levels should be taken into account during the design process of any technology solution focused on this group. We must also consider that older people often do not feel comfortable in handling a computer and the use of technological devices seems complicated for them. This problem may be worse considering the decline in cognitive, visual or motor abilities.

The Naviga project (2009-2012) is an European initiative funded by the Eurostar [2] R&D program and whose main objective is to provide these collective tools, devices and methods to enjoy personal autonomy and a better quality of life, to do that, within the project we are developing an integrated technology platform to provide Internet access through a computer or TV. In addition, the proposed platform will facilitate the incorporation of elderly's and people with different functional capacity to the Information Society through the use of special devices, social networks, and applications to improve the cognitive ability or personalized health services. The consortium comprises five SMEs conducting research (investment min. 20% of annual turnover in R & D), a university and two end users (an hospital and a daily care health centre located in Madrid region) are also involved in the project.

2 Objectives of Naviga Project

The Naviga project, through the use of information and communications technology, intended to cover a range of social and health objectives aimed to improving access to Information Society by the elderly's and people with disabilities. Within Naviga we will develop an open platform and adaptive technology for various purposes detailed in the following subsections.

2.1 Technological Oriented Objectives

On the one hand, the development of an adaptive communication interface between user and computer or television, to facilitate the understanding of Internet and new technologies to people with a low-tech profile, while encouraging its use by providing a simple and friendly human machine interface. Also, this interface takes into account the integration with different support products on the market to ensure that users can use those techniques. Furthermore, the development of a platform that allows rapid creation of services and applications specifically for the elderly's and disabled people with a common API.

2.2 Social Oriented Objectives

At the other hand, the main social objectives lies in the attempt to bridge the gap that prevents the elderly's and people with disabilities access the Information Society. To do this, we are developing simple mechanisms for interaction between technical elements (computer, television or special input devices), and people, for example an accessible Web browser to improve usability through the use of alternative hardware to keyboard as for example voice commands. Also, the browser will be compatible with common support and aid products for elder people, also we are developing a social network among people with the same disability, where users can find people with common interests and concerns, and share information, experiences and advices. An example would be evaluate and recommend support products, as these aids often have a high cost and does not respond equally to all profiles of disability.

2.3 Health Oriented Objectives

Similarly, the Project will provides a range of health-oriented goals that help elderly's to keep active through mental training exercises, and otherwise assist staff medical (hospitals, health centres) in monitoring the treatment of these people from homes, in this case developing services and games that allow mental training (mind training), suggesting exercises to keep the mind active, and getting people to communicate and participate to a greater extent in their social community. This will prevent premature degeneration of mental activity, and improves the mood of older people with functional diversity by increasing the feeling of being useful to society around them. Although little is known about the perceived benefits of digital games for the elderly, there is a small but growing body of research evidence in support of the notion that digital games can have a significant positive impact on the elderly`s mental and physical health and wellbeing [3]. Some research [4] has showed the benefits of gaming for elderly people in several domains: stimulation of social interaction and participation; enhancement of perceptual-motor skills (eye-hand coordination, dexterity, and fine motor abilities); improvement of performance speed (basic movements and reaction time); information processing, reading, comprehension, memory, self-image, etc. and transfer of the skills acquired in the games to other aspects of everyday routine like automobile driving.

Development of personalized health services is also part of the Naviga's objectives, such as warning and reminder system for medication adherence through an automatic smart pill dispenser or rehabilitation physiotherapy through virtual reality applications. In the latter case, the main goal is to recover the functionality of the hand of patients using a glove that makes measurements of the angles of each phalanxes up to 22 degrees of freedom with high accuracy. The device uses a strain sensing technology that transforms the movement of the hand and fingers to digital data in real time.

3 Architecture of Naviga Technology Platform

The technology platform being developed within the Naviga Project, see Figure 2, must solve two major technical challenges:

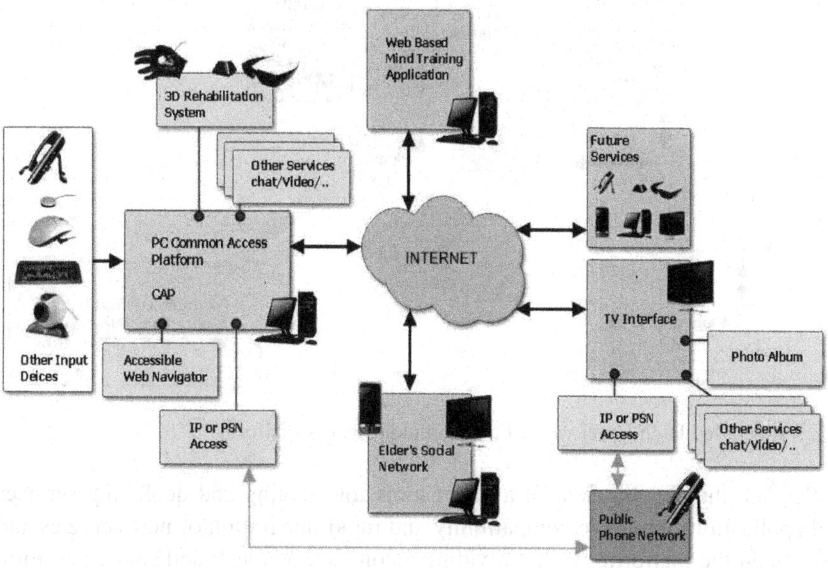

Fig. 2 Naviga's Architecture Technology Platform

Firstly, the connection to the platform in an interoperable way of different support products and communication interfaces, integrating health monitoring devices that generate medical alerts, fall detection systems and security alarms, and devices that enable accessibility to users with motor or cognitive disabilities to information and entertainment services, and advanced communications such as videoconferencing.

The number of support products available in the market is very high, but often not compatible with each other or have the same degree of utility to different users who share a disability. It is therefore necessary to develop a common multi-modal interface that simplifies the integration between computer and any specific support product. It should also be taken into account the need for multi-channel access, allowing Internet access through the computer, television or mobile devices.

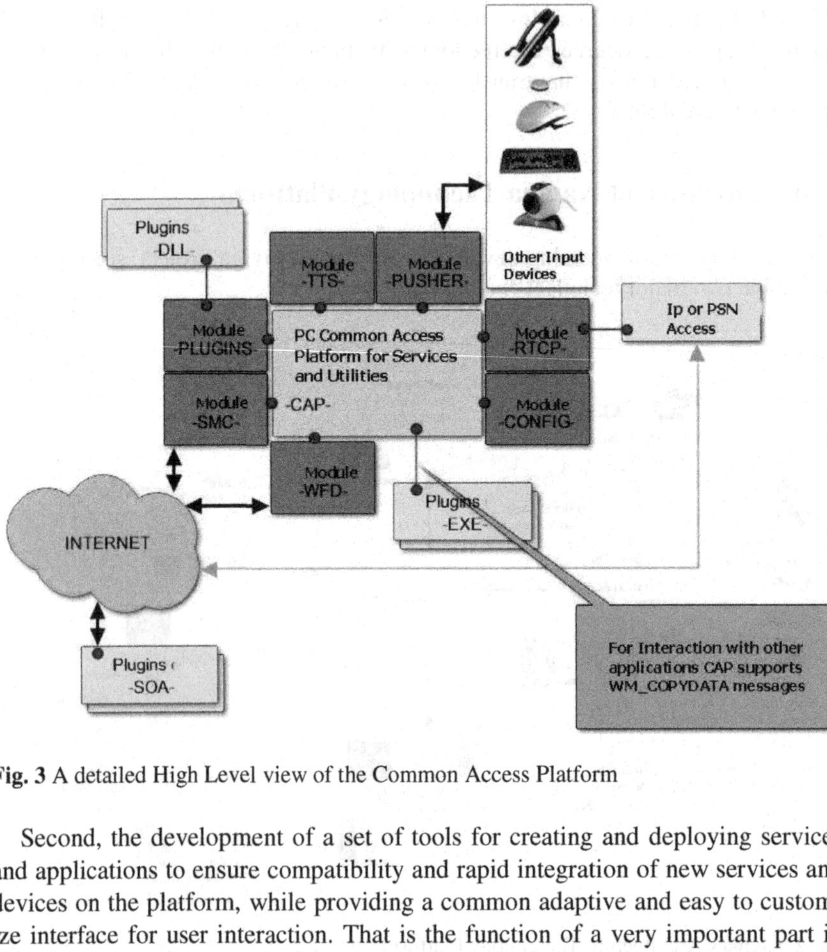

Fig. 3 A detailed High Level view of the Common Access Platform

Second, the development of a set of tools for creating and deploying services and applications to ensure compatibility and rapid integration of new services and devices on the platform, while providing a common adaptive and easy to customize interface for user interaction. That is the function of a very important part in the Naviga's technological architecture as is the Common Access Platform, see Figure 3, a shell running in place of the Operating System that implements several modules for Short Message Commands, Web File Download, Text-to Speech, etc. and all elements for manage future applications and services to be included in the Project.

The Naviga project aims to simplify the integration of new services and applications within the platform using freely available technologies that allow the subsequent adaptation of the code easily, so can still be used for further developments. Among the initial services of the platform, there are technical difficulties related to the application area. For example, the development of an accessible Web browser must be multimodal and interoperable in order to take into account the needs of all members of the group, which greatly complicates the solution given the diversity of users. Also, the use they make of the social network can be very different, both, use objectives (social relationships, share experiences, recommend support products) and access to services, must provide simple user interfaces, easy to use and highly adaptability to the preferences and characteristics of each person.

NAVIGA platform provides an open system based on SOA (Service Oriented Architecture) that enables and facilitates the development of new applications and services that seamlessly integrate with existing modules without need of an expert knowledge of the lower layers architectures and languages. Also, open source implementation based on Java EE and scripting languages like JavaScript, and compliance with accessibility standards of the ISO and the recommendations of the WAI, ensure continuity of service and support the development of the platform.

Designing the platform in conjunction with the devices ensures optimum performance and response to user actions, as adapted interaction mechanisms must play sometimes very complex tasks from very simple input actions. The end-user participation in the project to determine more accurately their needs and desired objectives, and prototype validation during the development process in aspects as interface usability and effectiveness of associated devices, verifying compliance with the requirements.

4 Expected Results

As mentioned above, from the point of view of development, the project's expected results are:

- A hardware interface device adaptable to all seniors and people with disabilities enabling the interaction with computer or television.
- A framework (tools and methods) for creating and deploying services and applications.
- The development of services including a Web browser that allows access for elderly's and disabled people to the Internet.
- Two technology demonstrators in the field of e-Health and entertainment.
- An analysis of business opportunities and business requirements (identifying their strengths and weaknesses) for the successful commercialization of project results.

During the running of Naviga project two case studies /scenarios will be implemented, to demonstrate the functionality of the framework developed. One dealing with rehabilitation at home based on virtual reality, while another scenario will be developed and evaluated in a care centre for elderly's and people with disabilities aiming their access to the Information Society through the Web browser and in particular social networks and mental training. The scenarios will have real participation of end users to validate the technological advances.

Acknowledgments. The R+D+i Project NAVIGA described in this paper is partially funded by the Center for Industrial Technological Development (CDTI) as part of the "Subprograma Interempresas Internacional" (CIIP-20091007).

References

1. Burdick, D.C., Kwon, S.: Gerotechnology: Research and Practice in Technology and Aging. Springer Publishing Company, New York (2010), 2004 program Eurostar website `http://www.eurostars-eureka.eu/`
2. Gamberini, L., Barresi, G., Mager, A., Scarpertta, F.: A Game of a Day Keeps the Doctor Away: A Short Review of Computer Games in Mental Healthcare. Journal of CyberTherapy and Rehabilitation 1(2), 127–145 (2008)
3. Griffiths, M.: The therapeutic value of video games. In: Raessens, J., Goldstein, J. (eds.) Handbook of Computer Games Studies, pp. 161–171. The MIT Press, Cambridge (2005)
4. RM-ODP Web site (2010), `http://www.rm-odp.net/`
5. Gartner, Gartner Reveals Five Social Software Predictions for 2010 and Beyond (2010), `http://www.gartner.com/it/page.jsp?id=1293114`
6. Griffiths, M.: The Therapeutic value of video games. In: Raessens, J., Goldstein, J. (eds.) Handbook of Computer Games Studies, pp. 161–171. The MIT Press, Cambridge (2005)
7. Jimison, H.B., Pavel, M., McKanna, J., Pavel, J.: Unobtrusive Monitoring of Computer Interactions to Detect Cognitive Status in Elders. IEEE Transactions on Information Technology in Biomedicine 8(3), 248–252 (2004)
8. Czaja, S.J., Lee, C.C.: Information technology and Older Adults. In: Jacko, J.A., Sears, A. (eds.) The Human-Computer Interaction Handbook, 2nd edn., Erlbaum, New York (2007)

A DSL for Context Quality Modeling in Context-Aware Applications

José R. Hoyos, Davy Preuveneers, Jesús J. García-Molina, and Yolande Berbers

Abstract. Developing reliable context-aware applications remains a big challenge, even after a decade of research in this area. Usually a lot of code is required to handle an application's correct behavior in a variety of different situations. Along with a growing amount of code, also increases the risk of programming errors that may lead to an undesired behavior in particular situations. In this paper we present a domain specific language (DSL) for developing context-aware applications. It allows creating context quality models which are transformed into software artifacts of the final application. This code generation saves time and effort, and helps to ensure an appropriate autonomic behavior at runtime in inherently uncertain situations.

Keywords: Domain Specific Language, Context-Aware, Quality, MDD, Context Quality.

1 Introduction

Developing adaptive and proactive context-aware [2] systems is a complex process. It is a big challenge to achieve a reliable behavior and quality of service (QoS) in a variety of circumstances. Not only the implementation, but the whole design phase should unambiguously describe how the system should operate in different situations. Software engineering techniques like Model Driven Development (MDD) can help to achieve this, using models at different levels of abstraction. These models

José R. Hoyos · Jesús J. García-Molina
Departamento de Informática y Sistemas
Universidad de Murcia, 30100 Murcia, Spain
e-mail: {jose.hoyos, jmolina}@um.es

Davy Preuveneers · Yolande Berbers
Department of Computer Science, Katholieke Universiteit Leuven
Celestijnenlaan 200A, B-3001 Heverlee, Belgium
e-mail: {davy.preuveneers, yolande.berbers}@cs.kuleuven.be

P. Novais et al. (Eds.): Ambient Intelligence - Software and Applications, AISC 92, pp. 41–49.
springerlink.com

can be expressed using well-defined languages called Domain Specific Languages (DSL) from which model transformations can generate code. This process saves time and effort to develop the system.

Pervasive computing is typically highly sensor-driven, but all too often so called smart context-aware systems wrongly assume that the information they retrieve from context sensors in the surrounding is accurate and relevant. Sensors are prone to noise and imperfections, and context information can be ambiguous. Having a notion of context quality helps to select reliable context sources, to be confident about a particular context situation, and to trigger appropriate proactive behavior.

The MDD approach has not been well exploited by the research community for context quality modeling. With most tools mainly focusing on the business logic of an application, it is very difficult to express specific context situations and context quality requirements to define and restrict an application's behavior.

MLContext is a DSL specially tailored to model context, which is part of an MDD framework we are developing for context-aware systems [4]. In this work we present an MLContext extension for context-quality modeling. With this extension, to the best of our knowledge, it is the first DSL which lets the developer not only model the context quality attributes but also the applications context quality requirements, keeping the context model details separated from the business logic details. Another contribution is the domain analysis we carried out to define the quality extension of MLContext. We have identified the most commonly used quality attributes in the literature. We will also show how they can be combined to compute context quality for aggregated contexts.

This paper is organized as follows. In section 2, a survey of current related work is presented. In section 3, the main issues related to the representation of context quality are analyzed in order to identify the elements for our DSL. In section 4, the DSL is presented by means of an example. We also explain how the context and quality models can be used for computing context quality. Section 5 shows automatic generation of code from these models. Finally, in section 6, we end this paper with conclusions and opportunities for future work.

2 Related Work

A UML-based structured model of context quality is presented in [10]. This approach provides quality parameters for sensor and context situations. It also includes an aggregation model which describes aggregation of quality from sensor level to context situations. These models are subjected to UML notation restrictions and they can not specify in detail the context situations. Their aggregation model can only specify which quality parameters are related to a context situation. This approach does not generate any code from the models. MLContext does not use a separated aggregation model, as this kind of aggregation is implicit in our quality model. This can be achieved by defining the quality requirements for each context

Table 1 Quality attributes classification

Data acquisition		Data representation		Data usage	
precision	accuracy	units	format	believability	relevance
location	timeliness	understandability	aliases	completeness	availability
coverage	range			comparability	

situation because they are related with the quality attributes of the sensors. Another approach for modeling context situations without quality requirements is described in [1]. The authors make a distinction between a model of the environment (which represents concepts that are relevant to the system and their relationships) and a model of context (which represents facts, made machine interpretable through this model). The first model is defined by using an ontology, and the second by using the Resource Description Framework (RDF). A context is represented in the form of a triple (e.g. <*Adrian, hasLocation, Room01*>). This kind of representation has a low level of abstraction, and the authors report difficulties to model some advanced relationships for constraints such as "close to", which can be easily expressed using MLContext.

3 Domain Analysis on Context Quality Modeling

When defining a DSL, a domain analysis is required to identify the main concepts of the language and the relationships between them. We have analyzed several existing approaches [6, 7, 8, 9] on context quality with the aim of extending MLContext with constructs for modeling and measuring quality in a context. These approaches organize quality parameters in several dimensions, but they do not consider that quality issues at low level (i.e. sensor data) are different from those at higher levels of context (i.e. relevance and ambiguity) [5].

In some cases we found an overlap of quality attributes belonging to different dimensions, and even a different interpretation for the same quality attributes (e.g. accuracy and precision). We classified in Table 1 the most common quality attributes based on three categories: *data acquisition*, *data representation* and *data usage*. This classification is simple and avoids overlap. Our context model can be extended to include new quality attributes.

- **Data acquisition**: these are directly related to the sensors that captured the information and must be defined at the sensor level.
- **Data representation**: these are used to specify acceptable representations of information that can be correctly interpreted.
- **Data usage**: these define the context of the data acquisition itself to decide if the data is relevant.

As a result of this domain analysis, we have identified the following concepts: *entity, source of context, provider, situation*. An *entity* could be any person, place, physical or abstract object, or event relevant for the interaction between a user and an application (cfr. Dey's definition of context [3]). The value of the properties of an entity can be obtained from a *source of context*. This context source is attached to a physical adapter or *provider*. A *situation* is a set of constraints on contextual facts, i.e. *context information*, so that when these constraints are satisfied in the sensed environment, the situation is said to occur [1]. For example, the health status of a patient could be composed of his temperature, blood pressure and pulse with *'the patient is ill'* being described with a constraint *'temperature* $>$ *37°C'*. The aim of this work is to compute the global quality of this context situation based on the quality of the pieces of context information it is composed of.

We have identified these requirements to extend MLContext:

1. The DSL must allow to express the following:
    - Quality attributes of context sources
    - Quality requirements for context-aware applications
    - Different context situations to support behavior variability in the design phase
    - Composition of complex context situations with aggregation of simple ones
2. The DSL must support inference of quality for aggregated contexts.

4 A DSL for Context-Quality Aware Applications

In this section we show how to model context quality using our DSL by means of a simple example. Let us suppose a user is in his car and he is trying to find an open gas station (e.g. in a 1 Km radius). There are four steps in our process of modeling: (1) Model entities, (2) Define sources of context, (3) Specify providers with quality attributes, and (4) Model context situations with quality requirements. We use the MLContext syntax for steps (1) and (2), and the MLContext quality extension presented in this paper for steps (3) and (4). The MLContext abstract syntax and concrete syntax is explained in [4].

Step 1: We will model the entities now. Our context model has three entities: the *user*, his *car* and the *gas_station*. The complex context of the *user* is composed of three simple contexts of *personal*, *environment* and *physical* type. The property "name" has a constant value "John" and is supplied by the user. The "location" property value can be obtained from the GPS of the *mobile_pda* source of context. The "time" property value can be obtained from the *mobile_pda* source but also from the system *clock*. The entity *car* has a simple context of environment type with a property "contains" expressing the fact that the *user* is in the *car*. The value of the "location" property of the *car* can be obtained from the *car_GPS* source of context. Finally, for the *gas_station* entity, we have modeled its "location", the "opening" and "closing" times. We used the *quality* modifier to specify that the location value is expressed in WGS84 format. The accuracy of "0" expresses an exact value.

```
contextModel example1 {
entity user {
    personal    { "name" :     "John" }
    environment { "location" source: mobile_pda }
    physical    { "time"     source: mobile_pda, clock } }
entity car {
    environment { "contains" : user
        "location" source: car_GPS } }
entity gas_station {
    environment { "location" : "38.0237616129431, -1.17448434233665"
                               quality: "units:WGS84, accuracy:0"}
    social { "opening_time" source: shops_system_service
             "closing_time" source: shops_system_service} } }
```

Step 2: Due to space constraints we only show the *mobile_pda* context source. The *interfaceID* keyword is a provider ID. The *methodName* statements are used to specify how the information is retrieved from the provider. "getLocation" and "getTime" are used to obtain the location of the user and the current time.

```
contextSource mobile_pda {
    interfaceID : "01-02-03-0A"
    methodName : "getLocation" {
            supply: user.location
            returnValue: "coordinates"}
    methodName : "getTime" {
            supply: user.time
            returnValue: "string"} }
```

The keyword *returnValue* specifies the type (format) of the returned value. The *car_GPS* and *clock* sources are defined in a similar way and are attached to providers "01-02-03-0B" and "01-02-03-0C" respectively.

We need another model for steps (3) and (4) to specify the quality attributes of the providers and the quality requirements for our application. This is an application dependent model, so following an MDD approach we define it separately from our context model.

Step 3: Each of the methods for supplying context information can have different quality attributes. Provider "01-02-03-0A" is linked to the *mobile_pda* context source. This provider can supply the location of the user (via GPS) and the current time, through the methods "getLocation" and "getTime". In this example, we have specified quality attributes for both methods. The location information is expressed in WGS84 units with a precision of 0.1E-12, but an accuracy of only 0.0064 (i.e. a 600m error approximately). The *timestamp* keyword indicates that this adapter can supply the time instant of the measurement. For the "getTime" method, we only specify that it provides the time using the SystemTimeDate units. Provider "01-02-03-0B" is attached to the *car_GPS* context source. It has the same precision as the *mobile_pda* but better accuracy (300m error approximately). The clock quality attributes are defined in a similar way but not shown here.

```
provider "01-02-03-0A" {
    location : user
    method : "getLocation" { units : "WGS84"     precision : "0.1E-12"
                             accuracy : "0.0064" timestamp }
    method : "getTime" { units : "SystemTimeDate" } }
provider "01-02-03-0B" {
    location : car
    method : "getLocation" { units : "WGS84"     precision : "0.1E-12"
                             accuracy : "0.0023" timestamp } }
```

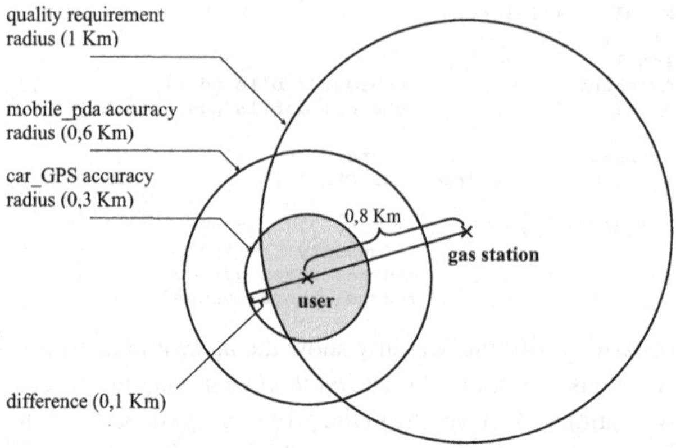

Fig. 1 Accuracy radius for the gas_station's location

Step 4: We can now express the quality requirements for our specific context-aware application. We are interested in finding a gas station near the user's location. Therefore, we can specify the following context situation:

```
situation open_gas_station_is_near {
  #user.location == #gas_station.location[accuracy="1", units="kilometers",
                                          relevance="1.2"]
  AND #user.time >= #gas_station.opening_time
  AND #user.time <= #gas_station.closing_time }
```

The *open_gas_station_is_near* situation is based on three context facts, which are specified by expressions concatenated by AND operators, so we compute the global quality for this situation as the product of the quality value of each of them. The first fact specifies that, if the user location and the gas station location are totally accurate, we accept as valid any reported locations which differ at most one kilometer. Of course, if the locations are not totally accurate, the quality of this context fact will not be 100%. *Relevance* is also an important subjective quality requirement. The developer must be able to express that certain context information is more important (relevant) than other ones and, therefore, its quality value has a higher contribution to the global quality of the context situation. In this example, we can suppose that it is more important to be sure about the distance from the gas station, because the user is interested in reaching the gas station (perhaps due to a low fuel level) even if he needs to wait until it is opened. The value of the *relevance* requirement is a factor which amplifies the maximum difference between the real value and the reported value of the location. By inspecting the models, the middleware knows it can obtain the value of the user's location property from the *mobile_pda* source. The user is in his car as may be inferred from the OWL-DL generated ontology, so the *car_GPS* can be added to the list of available sources for the user's location property. In this case, the middleware will choose the *car_GPS* source because it has better accuracy. Suppose the *car_GPS reports* a 0,8 Km distance from the *gas_station* (see Fig. 1).

Any user's location inside the 1 Km radius from the gas station will report a 100% of quality value but, there is a possibility that the user's actual position is outside the radius. The grayed zone of the accuracy circle for the *car_GPS* meets the requirements for the context situation we have defined. It represents a 83,33% of the total surface. Therefore, the quality value reported by the middleware for this fact might be 0,8333, but it is only 0,80 because of the relevance requirement.

The middleware can obtain the opening time for the *gas_station* from the *shops_ system_ service*. Let us suppose that this source provides the "10:00" value for the opening hour, with a 30 minutes accuracy. The current time (user.time) is "10:22", so there is an 8 minutes interval in which the *gas_station* might be closed. Therefore, the quality of this context fact is 86,67%. The quality value for the last context fact is 100% because current time is far from the closing time. The quality value reported for the global context situation will be 0,80 x 0,8667 x 1 = 0,6934.

5 Automatic Code Generation

The MLContext engine automatically generates software artifacts of the context-aware application from a context model. An OWL-DL ontology and Java classes for context providers were generated for the OCP middleware as described in [4]. Whereas the generation of these Java classes was specific to OCP, the ontology could be used by other middleware. In this paper, the ontology is used by the middleware to infer that the user is in his car and, therefore, the car GPS can be used to obtain his position. An OWL *class* is created for each of the categories in the model and a *DatatypeProperty* for each of their properties, taking into account the inheritance hierarchy. When a property refers to other entities (relationship), we create an OWL *ObjectProperty*. With regards to the MLContext quality model, a quality framework has been defined which can be used to provide functionality related to the quality management to any existing middleware. Fig. 2 shows a partial view of

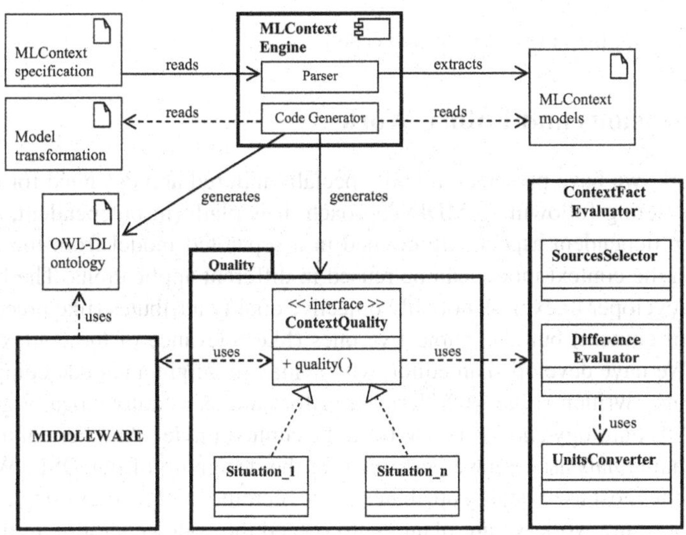

Fig. 2 Model-based framework to generate software artifacts for context quality

this framework and the generated artifacts by the MLContext engine. This engine automates the instantiation process by generating Java classes for each context situation. These classes implement the interface ContextQuality which has a quality method for computing the quality value of the situations based on the quality of their context facts. The evaluation of the quality of a context fact is performed by a *ContextFact Evaluator*. It uses a *Difference Evaluator* which calculates the maximum difference between the required value of a context property and the expected value from the context source. The framework also includes a UnitsConverter, which can change from some units to others, and a *SourceSelector* for selecting the (most suitable) best context source to obtain the value of a context property if it has more than one source. Fig. 3 shows an excerpt from the generated Java code for the example presented in this paper.

```
public class open gas station is near implements IContextQuality {
  private float CF1; //user.location == gas station.location
  private float CF2; //user.time >= gas station.opening time
  private float CF3; //user.time <= gas station.closing time

  public open gas station is_near () {
    ContextFactEvaluator CFE = new ContextFactEvaluator ();
    List<QualityReq> LOFQR1 = new ArrayList<QualityReq> ();
    LOFQR1.add(new QualityReq("accuracy",QualityToolsPackage.OPEQUAL,"1"));
    LOFQR1.add(new QualityReq("units",QualityToolsPackage.OPEQUAL,"kilometers"));
    LOFQR1.add(new QualityReq("relevance",QualityToolsPackage.OPEQUAL,"1.2"));
    CF1=CFE.evaluate("user.location",
                     "gas_station.location",QualityToolsPackage.OPEQUAL,LOFQR1);
    CF2=CFE.evaluate("user.time",
                     "gas_station.opening time",QualityToolsPackage.OPGREATE,null);
    CF3=CFE.evaluate("user.time",
                     "gas_station.closing time",QualityToolsPackage.OPLESSE,null);
  }
  public float quality () {
    float QualityValue;
    QualityValue=CF1*CF2*CF3;
    return QualityValue;
  }}
```

Fig. 3 An excerpt from the generated Java code

6 Conclusions and Future Work

In this work, we have presented a DSL specially tailored and designed for context quality modeling following a MDD approach. It is platform independent, and the application dependent aspects are defined in a separated model from the context aspects, so the context model can be reused in different applications. The DSL allows the developer to express not only objective quality attributes (like precision of a source of context) but also subjective ones (like relevance of the context information). We have developed an editor, with syntax highlight and code completion. We have also written some MOFScript transformations to automatically generate an OWL-DL ontology and Java code from the context model. Another contribution is the domain analysis we have performed in the designing of our DSL. We have identified the most used quality attributes in the literature and their existing relationships. As a future work we are planning to extend the code generation to different middlewares and to use the context model for predicting context situations.

References

1. Clear, A.K., Dobson, S., Nixon, P.: An approach to dealing with uncertainty in context aware pervasive systems. In: UK/IE IEEE SMC Cybernetic Systems Conference 2007, Dublin, IE. IEEE Press, Los Alamitos (2007)
2. Dey, A.K.: Understanding and using context. Personal Ubiquitous Comput. 5(1), 4–7 (2001)
3. Dey, A.K., Abowd, G.D.: Towards a better understanding of context and context-awareness. In: Proceedings of the Workshop on the What, Who, where, when and How of Context-Awareness, ACM Press, New York (2000)
4. Hoyos, J.R., García-Molina, J., Botia, J.A.: MLContext: A Context-Modeling Language for Context-Aware Systems. In: 3rd DisCoTec Workshop on Context-aware Adaptation Mechanisms for Pervasive and Ubiquitous Services, vol. 28 (2010)
5. McKeever, S., Ye, J., Coyle, L., Dobson, S.: A multilayered uncertainty model for context aware systems. In: proceedings of the International Conference on Pervasive Computing: Late Breaking Result, pp. 1–4 (May 2008)
6. Strong, D.M., Lee, Y.W., Wang, R.Y.: Data quality in context. ACM Commun. 40(5), 103–110 (1997)
7. Wand, Y., Wang, R.Y.: Anchoring data quality dimensions in ontological foundations. ACM Commun. 39(11), 86–95 (1996)
8. Wang, R.Y., Strong, D.M.: Beyond accuracy: what data quality means to data consumers. J. Manage. Inf. Syst. 12(4), 5–33 (1996)
9. Watts, S., Shankaranarayanan, G., Even, A.: Data quality assessment in context: A cognitive perspective. Decis. Support Syst. 48(1), 202–211 (2009)
10. Ye, J., McKeever, S., Coyle, L., Neely, S., Dobson, S.: Resolving uncertainty in context integration and abstraction: context integration and abstraction. In: ICPS 2008: Proceedings of the 5th International Conference on Pervasive Services, pp. 131–140. ACM, New York (2008)

Non-intrusive Residential Electrical Consumption Traces

Marisa Figueiredo, Ana de Almeida, and Bernardete Ribeiro

Abstract. An active trend of research consists of understanding how electricity is used, namely for energy efficiency enforcement and in-home activity tracking. The obvious and cheapest solution is to use an inconspicuous monitoring system. Through the use of non-intrusive load monitoring systems, the signal from the aggregate consumption is captured, electrical significant features are extracted and classified and the appliances that were consuming are identified. In order to obtain a precise identification of the device, the main requirements are an electrical signature for each device and a proper classification method. The information thus obtained identifies appliance's usage and specific consumptions. This paper describes an on-going research aiming at the development and simplification of techniques and algorithms for non-intrusive load monitoring systems (NILM). The first steps in the implementation of a NILM system were already addressed, namely we are performing an extensive study on the characterization of the appliance's electrical signature. The proposed parameters for defining an efficient electrical signature are the step-changes in the active and reactive power and the power factor.

Keywords: Activity modelling, Appliance Identification, Feature Extraction, NILM.

1 Introdution

In March 2007, the European Member States agreed on three common objectives for tackling today's energy and climate challenges until 2020, the so-called '3x20'. By 2020, the European Union should: reduce greenhouse gas emissions by 20%; reduce energy use by 20%; achieve 20% of renewable energy in its overall energy

Marisa Figueiredo · Ana de Almeida · Bernardete Ribeiro
CISUC, Department of Informatics Engineering,
University of Coimbra, Polo II, P-3030-290 Coimbra, Portugal
e-mail: mbfig@dei.uc.pt, amaria@dei.uc.pt, bribeiro@dei.uc.pt

P. Novais et al. (Eds.): Ambient Intelligence - Software and Applications, AISC 92, pp. 51–58.
springerlink.com

supply. According to the European Environment Agency [18] the largest energy consumption sector is the household sector (29% of the overall consumed energy). Alerting the consumers of their true consumption behavior is a key action issue for achieving the 3x20 targets. This means that end users must be aware not only of the overall load consumption but also of how is the energy being spent. Such information is also very useful for recent concepts such as Smart Grids and even more so in-Home Activity Tracking, where an accurate and inconspicuous monitoring of electrical appliances consumptions is required.

Currently the market offers load consumption monitoring solutions such as smart meters and individual meters. A smart meter is a sensing device that supplies data about the aggregated consumption without identifying which particular devices are on. This limitation can be overcome by using an individual meter for each appliance in the household. Nevertheless, at the moment, this turns out to be an extremely costly and intrusive solution. A more viable solution for the monitoring of individual electrical devices is a non-intrusive load monitoring system (NILM). A single device is used to monitor the electrical system and to identify the loads of each appliance without needing extra sub-measurements [14].

The main goal of a NILM system is to identify which are the appliances switched on at a certain moment in time. Several works have shown that individual loads can be detected and separated from sampling of the power at a single point (e.g. the electrical service entrance for the house). Building such a system presents several challenges: the definition of electrical signatures for the appliances, the disambiguation of the aggregate signal (feature extraction) and device identification (classification). This data can be used for tracking of in-home activities (e.g. in order to capture changes in the behavior and issue alerts for Alzheimer's patients).

This paper presents an on-going research project that aims at the development and simplification of techniques and algorithms for NILM systems. The first steps in the implementation of a NILM system were already addressed, namely we are performing an extensive study on the characterization of the appliance's electrical signature. The proposed parameters for defining an efficient electrical signature are the step-changes in the active and reactive power and the power factor. The paper is organized as follows: the next section describes the concepts behind a NILM system and the electrical identification problem. In Section 3 a brief overview of the related literature is presented. The following section describes the proposed approach and the main research questions that are being addressed. Some of the results already achieved are presented in Section 5. Section 6 concludes the paper.

2 Problem

The NILM concept can be described as a methodology which requires only a single device to monitor the functioning of several electrical appliances by recognizing the electric load related to each device. This must be achieved without increasing the marginal cost of electricity and without the need for sub-measurements in each appliance. This solution, in general, consists in three steps: signal sampling (current and voltage signals), feature extraction (information as active and reactive power are extracted from the sampled signals) and appliance identification (classification methods are used to identify the correspondent device) (see Figure 1).

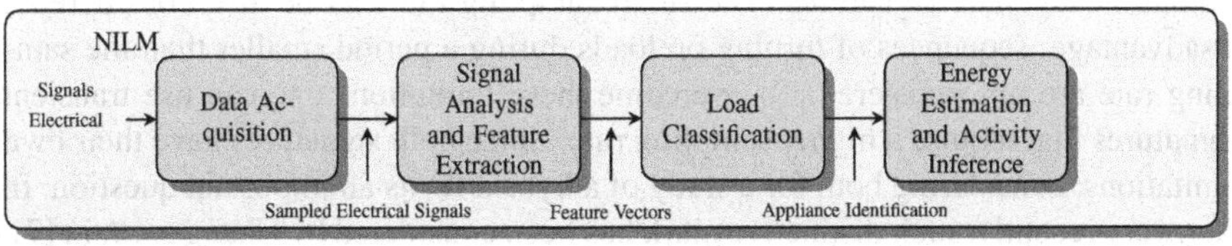

Fig. 1 A NILM high-level system

There are two relevant issues related to a NILM system: privacy and accuracy. The first arises from the fact that the NILM allows for the construction of consumption patterns of a household. This is a sensible subject since, like all personal data, this information can easily be misused (e.g., by the electric utility companies for marketing campaigns). The accuracy issue, that will be our focus, concerns to the correct identification of appliances when two or more of them are connected to the same circuit. The task of developing a system as accurate as possible is a complex one. This is mainly due to the similarity between device signatures, which so far is dependent on the sensing meter used to collect data. This means that different sensors obtain different value measurements. Nevertheless, general patterns seems to exist even using different measurement devices [4].

The information provided by a system with these characteristics can be used by the consumer and/or by a utility company. On the one hand, detailed information - about which are the devices used and when - can be feed backed to the consumer. On the other hand, this information can be used to trace electrical consumption profiles which is useful in a number of research areas.

3 Related Literature

The NILM concept was introduced by Hart [14] (Electric Power Research Institute) and by Sultamen (Electricité de France) [26], independently. Over the last decades, due to the pressing environmental and economic issues, the interest in this area has increased and the general theme was the focus of PhD theses as the one in [17]. In 1996, the first NILM system was commercialized by the company Enetics, Inc.

In order to accurately identify the appliances related to the occurring events, an electrical appliance ID is necessary. Two types of non-intrusive signatures were proposed by Hart in [14]: the steady-state and the transient signatures. The first are deduced from the difference between value states in the signal. The latter ones are based on the disturbance detected on the electrical signal by the turning on or off of an appliance (or the noise induced in the signal by these events). To perform steady-state detection is much less demanding than the capture and analysis of the transients. Other advantages are the fact that we can recognize turning off states and even perceive when two different appliances are on at the same time. Due to their advantages, steady-state signatures were used by several authors [14, 10, 5, 6] mainly for residential load monitoring systems. Nevertheless, some limitations can be pointed out as the impossibility to distinguish two different appliances with very similar steady-state signatures. The small sampling rate can be also considered a disadvantage: sequences of turning on loads during a period smaller than the sampling rate are not registered. To overcome these limitations one can use transient signatures that require a higher sampling rate. Since both signatures have their own limitations, considering both for a study of a hybrid ID its an important question. In fact, very recently, such distinctive mark has been considered by Chang *et al.* in [7].

Beyond the electrical devices signatures, signal analysis and classification methods are a major part of the system. Methods as *1-Nearest Neighbor*, *Decision Trees* [5], *Least square criterion* [16] and *Artificial Neural Networks* [8] are some examples of methods that have been employed to perform load classification. A summarized state of the art on load monitoring methods can be found in [19] where a comparison between the steady-state, harmonic and transient analysis is presented.

The inference of in-Home activities from the results of electrical load disambiguation is most useful for the "Ambient Assisted Living" research area. However, only few works were found in the related literature. The work found in [3] proposes an approach to detect activities of daily living, using as input the data provided by the Watteco sensor. In fact, the existence of a NILM system can lead to the development of healthcare products and electricity consumption profiles. However, the potential of the NILM system is not restricted to these areas. Others feasible applications of the data provided by a NILM technique are: fault detection, monitor of HVAC air filters [23, 20], and shipboard systems [25].

4 Research Objectives and Approach

The basic motivation for this research project is the study and prototyping of an NILM system allowing for the identification of each device from the aggregated electrical load. Currently, the strongest shortcoming comes from the difficulty in obtaining an accurate identification when more than one device with similar electrical signatures are on. With this research, answers to questions like "Does the unit signature reach good identification accuracy?", "How to deal with intermediate states of some appliances?" and "How to handle with the simultaneous events?" are intended to be studied. We investigate the possibility of overcoming this limitation by using transient signatures. Therefore, this is the first research question that must be answered in order to produce a more consistent electrical appliance ID. The study starts with the analysis of both types of signatures separately. For this, the signal needs to be sampled at a higher rate than usual and use existing databases in order to extract the relevant features. To analyze the transient signal, some authors have used the Fourier transform [21]. The Fourier transform is used to analyze stationary signals, where the frequency content does not vary with time. However, the transient signal is a non-stationary signal which does not have a constant frequency along the time, hence, if the time information is needed. Therefore, the Fourier transform is not a feasible solution. To overcome this limitation, it is expected to apply wavelets transforms to the transient signal since it is the more appropriate [2].

The next research step focus on the disambiguation of the aggregated signal. The most adequate methods for this particular feature extraction context will be studied. The features classification will be the obvious continuation. Again, after a throughout study of the existing classification methods the most adequate ones will be selected and adapted to this end. At present, the most promising one seems to be the "Support Vector Machines" [24], the "Clustering by Compression" method [9] (which has not yet been applied under this context). The most interesting fact about this recent method is that it does not need previous information about the data, as was showed by its authors in [9]. This is mainly due to the fact it is based on a formal measure of similarity between files. It was applied to a range of areas such as music and classification of fetal EEG signals [22] with considerable success.

This project main system is a tool to assist the iTEAM project team [1]. The iTEAM project intends to build a framework to evaluate the impact of the "green policies". To achieve this goal the iTeam will develop a simulator using individual/household and organization/firm agents. Therefore, the development of behavior models for the energy consumption, able to define activity profiles of the consumers, is proposed. The input of the activity model will be provided by the output from the NILM system.

5 Results

The study had to start by the formal definition and analysis of the steady-state signatures since they are less demanding in terms of sampling rate and therefore are the most used ones. A set of sequential samples will be identified as representing a steady-state if the distance between any two samples of the set is less than a given tolerance value. This threshold depends on the input signal and on the appliance in study. Other variable parameter is the minimum number of samples needed for a signal segment to be considered as a steady-state.

Steady-state signatures are deduced from the difference between states in the signal. In [12] we propose a technique to analyze the step-changes in a signal for the electrical signatures identification. The presented method is based on a relaxation and comparison of the areas on a step-wise electrical signal sampling. From the original proposal, a simplification was obtained by keeping only the extreme values already in the stable state and testing the new sample value against the previous ones. A paper presenting such an improvement was submitted to a conference [11]. The method starts be considering a sequence of n consecutive sampling values, $Y = \{y_i, i = 1, \ldots, n\}$ already identified as a steady-state. By definition, $|y_i - y_j| \leq \varepsilon \ \forall i, j = 1, \ldots, n$ and $i \neq j$, where $\varepsilon > 0$ is the defined tolerance. Let $y_M = \max\{y_i\}$ and $y_m = \min\{y_i\}, \forall i = l, \ldots, n$ be the maximum and minimum values, respectively, for Y, and that y_r $(r = n+1)$ is the next sample value. We proved that y_r maintains the stable behavior of Y if $y_M - \varepsilon \leq y_r \leq y_m + \varepsilon$ is satisfied.

At this point some related concepts must be introduced: active power, reactive power and power factor. In a simple alternating current circuit, current and voltage are sinusoidal waves that, according to the load in the circuit, can be in phase or not. For a resistive load the two curves will be in phase and multiplying their values, at each instant, produces a positive result. This indicates that only real power is transferred in the circuit. In case of a reactive load, current and voltage will be ninety degrees out of phase, which suggests that only reactive energy exists. In practice, resistance, inductance and capacitance loads will occur, so both real and reactive power will exist. At last, the product of the root-mean-square voltage values and current values represents the apparent power. The real, the reactive and the apparent powers are measured in Watts, volt-amperes reactive (VAR) and volt-ampere (VA), respectively. The relation between the three parameters is given by $S = P^2 + Q^2$ where S, P and Q represent the apparent power, the active and the reactive powers, severally. The apparent and the real powers are also connected by the power factor. The latter constitutes an efficiency measure of a power distribution system and can be computed as the ratio between real power and apparent power in a circuit. This ratio has values ranging in $[0,1]$ (0 for purely reactive load and 1 for resistive load).

The above described method was applied in the analysis and recognition of steady-states occurring in the active and reactive power signals and the power factor measurements. For each one of the different appliances tested was possible to identify three different steady-states in the signal: one previous to the switch on of any of the devices; one other corresponding to the appliances' operation phase and a last one occurring after switching off. In fact, one LCD in particular presented four different states since it was possible to identify the steady-state related to the LCD's standby mode [11].

The features extracted were used as an electrical distinctive characterization of appliances from a signature composed by information of active, reactive powers and power factor. To evaluate the effectiveness of this signature, data from a set of appliances were collected and classified using SVM and 5-NN classification one-against-all tests, as well as SVM multi-class [15] classification tests. In average, an accuracy around 96% for the Linear SVM, 97% for the RBF SVM and 99% for the 5-NN were obtained. Rather, the SVM multi-class presented very low accuracy values, around 40%. These results, reported in [11], indicate that an accurate identification of the devices can be, in fact, accomplished.

The present phase consists of collecting data in order to perform more ambitious tests, including the major test that consists of using several appliances switched on simultaneously. Only after completing this study will the transient signatures be studied in the same systematic analysis. Finally, the integration of both approaches in a single hybrid signature will take place and its effectiveness will be accessed.

The planned progress of the project can be impaired by the availability of specialized equipment to acquire the transient information. In order to do so, as stated above, the signal needs to be sampled at a high rate. The use of existing databases in order to extract the relevant features, to proceed with the study, can be a possible way to overcome this shortcoming.

6 Conclusions

The requirements of automatic solutions for the identification of electrical appliances can be fulfilled by a NILM system, which can also be used in a household electrical management system. This type of system is based on the definition of the electrical signature for each device, which must allow to achieve an accurate identification. The research project here described aims to achieve this goal by proposing a hybrid signature. This electrical ID must be precise enough to allow for a computationally simple enough implementation in order to be effective. Despite of the fact that the central goal to be the most accurate identification of appliances, there are other satellite questions related to privacy issues and hardware solutions. The output data of a NILM system can be used for several areas, from energy management systems to marketing campaigns of utility companies or even to residential surveillance[13]. Other open issue is related with the hardware associated to the implementation of an innovative NILM system. For instance, the NILM solution software can be either embodied on a non-complex and physically small device system or run on a computer unnoticed, using mobile technologies to acquire the data.

Acknowledgements. The authors would like to thank iTeam project for the grant support given.

References

1. Almeida, A.M., Ben-Akiva, M., Pereira, F.C., Ghauche, A., Guevara, C.A., Niza, S., Zegras, C.: A framework for integrated modeling of urban systems. In: 45 ISOCARP Congress 2009 (2009)
2. Angrisani, L., Daponte, P., D'Apuzzo, M.: Wavelet network-based detection and classification of transients. IEEE Transactions on Instrumentation and Measurement 50(5), 1425–1435 (2001)
3. Berenguer, M., Giordani, M., Giraud-By, F., Noury, N.: Automatic detection of activities of daily living from detecting and classifying electrical events on the residential power line. In: 10th International Conf. on e-health Networking, Applications and Services, pp. 29–32 (2008)
4. Berges, M., Goldman, E., Scott Mattews, H., Soibelman, L.: Training load monitoring algorithms on highly sub-metered home electricity consumption data. Tsinghua Science and Technology 13(S1), 406–411 (2008)
5. Berges, M., Goldman, E., Scott Matthews, H., Soibelman, L.: Learning systems for electric consumption of buildings. In: ASCE International Workshop on Computing in Civil Engineering, Austin, Texas (2009)
6. Bijker, A.J., Xia, X., Zhang, J.: Active power residential non-intrusive appliance load monitoring system. In: IEEE AFRICON (2009)
7. Chang, H.-H., Lin, C.-L., Lee, J.-K.: Load identification in nonintrusive load monitoring using steady-state and turn-on transient energy algorithms. In: 14th Intl. Conf. on Computer Supported Cooperative Work in Design, pp. 27–32 (2010)
8. Chang, H.-H., Lin, C.-L., Yang, H.-T.: Load recognition for different loads with the same real power and reactive power in a non-intrusive load-monitoring system. In: 12th International Conf. on Computer Supported Cooperative Work in Design, pp. 1122–1127 (2008)

9. Cilibrasi, R., Vitnyi, P.M.B.: Clustering by compression. IEEE Transactions on Information Theory 51, 1523–1545 (2005)
10. Cole, A.I., Albick, A.: Algorithm for non intrusive identification of residential appliances. In: Proc. of the 1998 IEEE Intl. Symposium on Circuits and Systems, ISCAS 1998, vol. 3, pp. 338–341 (1998)
11. Figueiredo, M., de Almeida, A., Ribeiro, B.: An experimental study on electrical signature identification of non-intrusive load monitoring (nilm) systems. Accepted by ICANNGA 2011 (2011)
12. Figueiredo, M., de Almeida, A., Ribeiro, B., Martins, A.: Extracting features from an electrical signal of a non-intrusive load monitoring system. In: Fyfe, C., Tino, P., Charles, D., Garcia-Osorio, C., Yin, H. (eds.) IDEAL 2010. LNCS, vol. 6283, pp. 210–217. Springer, Heidelberg (2010)
13. Hart, G.W.: Residential energy monitoring and computerized surveillance via utility power flows. IEEE Technology and Society Magazine (1989)
14. Hart, G.W.: Nonintrusive appliance load monitoring. Proc. of the IEEE 80, 1870–1891 (1992)
15. Joachims, T.: Making large-scale svm learning practical. In: Schlkopf, B., Burges, C., Smola, A. (eds.) Advances in Kernel Methods - Support Vector Learning. MIT Press, Cambridge (1999)
16. Laughman, C., Lee, K., Cox, R., Shaw, S., Leeb, S., Norford, L., Armstrong, P.: Power signature analysis. IEEE Power and Energy Magazine 1(2), 56–63 (2003)
17. Leeb, S.B.: A conjoint pattern recognition approach to nonintrusive load monitoring. PhD thesis, Massachusetts Institute of Technology (1993)
18. European Environment Agency. Energy and environment report. Technical report, European Union (2008)
19. Najmeddine, H., El Khamlichi Drissi, K., Pasquier, C., Faure, C., Kerroum, K., Diop, A., Jouannet, T., Michou, M.: State of art on load monitoring methods. In: IEEE 2nd Intl. Power and Energy Conf., PECon 2008, pp. 1256–1258 (2008)
20. Orji, U.A., Remscrim, Z., Laughman, C., Leeb, S.B., Wichakool, W., Schantz, C., Cox, R., Paris, J., Kirtley, J.L., Norford, L.K.: Fault detection and diagnostics for non-intrusive monitoring using motor harmonics. In: Applied Power Electronics Conf. and Exposition (APEC), pp. 1547–1554 (2010)
21. Patel, S.N., Robertson, T., Kientz, J.A., Reynolds, M.S., Abowd, G.D.: At the flick of a switch: Detecting and classifying unique electrical events on the residential power line. In: Krumm, J., Abowd, G.D., Seneviratne, A., Strang, T. (eds.) UbiComp 2007. LNCS, vol. 4717, pp. 271–288. Springer, Heidelberg (2007)
22. Costa Santos, C., Bernardes, J., Vitanyi, P.M.B., Antunes, L.: Clustering fetal heart rate tracings by compression. In: CBMS 2006: Proc. of the 19th IEEE Symposium on Computer-Based Medical Systems, pp. 685–690. IEEE Computer Society, Los Alamitos (2006)
23. Sawyer, R., Anderson, J., Foulks, E., Troxler, J., Cox, R.: Creating low-cost energy-management systems for homes using non-intrusive energy monitoring devices. In: Energy Conversion Congress and Exposition, ECCE 2009, pp. 3239–3246 (2009)
24. Schölkopf, B., Burges, C.J.C., Smola, A.J. (eds.): Advances in kernel methods: support vector learning. MIT Press, Cambridge (1999)
25. Shrestha, A., Foulks, E.L., Cox, R.W.: Dynamic load shedding for shipboard power systems using the non-intrusive load monitor. In: Electric Ship Technologies Symposium, ESTS 2009, pp. 412–419 (2009)
26. Sultanem, F.: Using appliance signatures for monitoring residential loads at meter panel level. IEEE Transactions on Power Delivery 6, 1380–1385 (1991)

Social Presence in Immersive 3D Virtual Learning Environments

Minjuan Wang, Sabine Lawless-Reljic, Marc Davies, and Victor Callaghan

Abstract. Ambient Intelligence (AmI) has historically addressed the interaction of people with computer controlled physical worlds. More recently, there has been interest in their virtual counterparts, such as SecondLife in which humanoid avatars interact with each other and their worlds in ways that are analogous to our relationship with the physical world. Virtual and physical worlds can have complex relationships ranging from either each augmenting the other, to the provision of services, such as eLearning. Virtual-Reality extends eLearning environments (eg regular audio, video, and text,) by enabling abstract concepts and entities to be given tangible forms within the virtual world. Also, students and teachers take the form of avatars allowing them to employ avatars to establish their social presence in a wide variety of ways. This paper introduces two popular virtual reality tools, presents a comprehensive review of the literature related to social presence and describes our practical work in progress towards constructing a mixed reality iClassroom.

Keywords: social presence, virtual learning environments, Second Life.

1 Introduction

Ambient Intelligence (AmI) describes environments in which intelligent computer processes mediate and enhance the interaction between people and technology. Initially the focus was on controlling physical environments but more recently there has been interest in their virtual counterparts, such as SecondLife which are mirrors of the physical world based on 3D dimensional *multiple-user virtual*

Minjuan Wang · Sabine Lawless-Reljic
San Diego State University & University of San Diego (U.S.)
e-mail: mwang@mail.sdsu.edu, sabinereljic@yahoo.com

Marc Davies · Victor Callaghan
University of Essex (U.K.)
e-mail: midavi@essex.ac.uk, vic@essex.ac.uk

P. Novais et al. (Eds.): Ambient Intelligence - Software and Applications, AISC 92, pp. 59–67.
springerlink.com

environments (MUVE). Such online immersive systems have their roots in earlier simulation and games technology and represent the convergence of the internet, social networking, simulation, and online gaming. Online applications are becoming increasingly social, offering multiple-participant options and social implications as typified by social networking games (e.g. World of Warcraft and Final Fantasy) and smart classrooms (e.g. the SJTU "Natural Classroom" [1]). The growing importance of online virtual environments is illustrated by reports such as that from Gartner, Inc. analysts [2], which predicted that "80% of internet users will be active in a virtual world by the end of 2011" and by the Pew Foundation which reported that "97% of teens play computer, web, portable, or console games" and that their gaming experiences "include significant social interaction and civic engagement" [3].

Another trend in virtual worlds is to provide more intelligent and engaging characters and environments together with more social iteration and an increase in user-generated content, effectively strengthening the users' sense of ownership and belonging in the environment [4].

2 The Technological Tools - Virtual Environments

In our work we use Second Life, and its derivative, RealXtend, as our virtual environment. However, these were not the first 3D social virtual worlds to be launched. Active Worlds came online in 1997 and quickly developed Active Worlds Educational Universe (AWEDU) to support its growing educational community. More recently, with the arrival of free open-source virtual world construction toolkits such as RealXtend, Unity 3D, Open Cobalt, OpenSim, Wonderland and Metaplace there is a growing number of virtual worlds that have been developed by individuals. In this paper we focus on Second Life and RealXtend (derived from Second Life), that we will introduce in the following sections.

2.1 Second Life

Second Life® (SL), launched in 2003, is a three-dimensional virtual world which, as of July 2009, had about 19 million users. The SL platform can be regarded as a merger of social networking tools (eg MySpace and Facebook) and online massively multi-player video games technology (egNeverwinter Nights, the first truly graphical multi-user role-playing games introduced in 1991 or EverQuest, released in 1999, and credited for introducing massively multiple-user online role-playing games mainstream to the West).

Second Life was created using the Linden Scripting Language (LSL), a scripting language similar to C. In January 2007, Linden Lab opened the source code to its client software, which runs on Windows, Mac OS X, and Linux. It is this client that users download on their PC to log into the virtual world of SL (for system requirements, see http://secondlife.com/support/system-requirements). SL exists on server farms which use Intel and Advanced Micro Devices computers located in geographically distributed facilities in the USA. Linden Lab' servers run Debian Linux and the MySQL database.

SL is a simulation based on physical metaphors of virtual worlds on virtual lands (geographic areas). Each geographic area represents a 256x256 meter region which runs as a single instantiation of a software process, called a simulator or "sim.. As the user's avatar moves through the world, it is handled off from one simulator process to another. When users buy or rent virtual lands in SL, they are indeed leasing software resource on a server.

Although used as a game by some users, SL was not built as a game platform, rather as a social interaction environment. SL, like Active Worlds and China's HiPiHi, is a socially-driven system. Although some scholars may disagree, Meadows has argued that users enter a completely metaphor-free environment in which rules are emergent and roles are entirely social [5]. Following the current gaming trend to increase content authoring by users, SL allows players to create content with user interface (UI) tools and modification software, and even 'hack' certain aspects of the platform's operating system to modify media and architecture, and the behavior of the avatars.

In order to *be* in SL, users must create a 3D alter ego called an avatar. Once logged in, the user has access to a UI that gives the avatar a rich sense of presence 'in world,' in the sense that SL allows people to interact via several senses such as creation of objects and landscapes, the manipulation of their appearance and behaviors and a rich array of communication modes between user/avatars including text, speech and avatar body language. This paper explores the effects of avatar social presences in the context of distance education.

2.2 RealXtend

For the practical development of our mixed reality*iClass*we use*RealXtend* [6]. This is open-source virtual world software, (programmed mainly using C# and Python). Its derived from the *OpenSim* (a.k.a. *Open Simulator*) project [7] which was based on code from *Second Life* [8] which allows us to benefit from the detailed graphics (i.e. realistic avatars and landscaping) of the popular online virtual world. As *RealXtend* is open-source we have complete access to modify any part of the software code. While *RealXtend* by itself solved our problems for landscaping worlds and avatars, it didn't contain models for creating realistic three-dimensional objects. To overcome these issues we turned to two free services provided by *Google*, specifically the *Google SketchUp 7* graphics editing suite, and *Google 3D Warehouse*, a vast online repository of three-dimensional models created by people using *SketchUp*, most of which are also free to use [9].

2.3 A Mixed Reality Intelligent Environment

Previous research applied AmI to creating smart-classrooms which have both a physical and online form [10]. Interaction both in, and between these environments can be undertaken using what is labeled mixed-reality. Using *RealXtend* we designed and built a three-dimensional virtual world to act as half of a mixed reality iClassroom the other half being the physical iClassroom. The physical iClassroom contained numerous hollow walls and ceilings which were outfitted with a myriad of embedded-computer based technologies, including both sensors and effectors. All of the technologies used in the iClassrom are wrapped into a generic OSGI UPnP framework. *RealXtend*can be augmented through the addition of Python scripts to add advanced features into the default world. Most of these function by being attached to specific prims (objects) in the virtual world. Whenever a prim is interacted with it then automatically runs the attached script code associated with the interaction method. Several Python scripts were created, including a bridge allowing real-world X-10 enabled devices (wrapped into a OSGI UPnP framework) to be remotely controlled using virtual counterparts and vice versa.

As *RealXtend* is based on *Second Life* the software has inherited the multi-user properties of an online virtual world, which have been passed on to the iClassroom's virtual environment. By allocating each online student with their own client avatar the *iClassroom* can be simultaneously inhabited by a collection of local and remote students, both of whom can interact with the class from anywhere in the world (via a computer or smart-phone with an internet connection), potentially allowing users from multiple age-groups and/or culture access to the environment.

3 The Pedagogical Issues - Virtual Reality and Social Presence

Previous research has emphasized the importance of presence in face-to-face education and of instructor presence in distance education. Virtual Reality goes some way to enabling presence as it enables users to engage in mediated social interaction including a full range of social interaction and contacts [9]. The popularity of SL has inspired many colleges and universities to explore usage of SL for hybrid and distance education, although, to-date, there has been very little research to justify the adoption and many questions remain unanswered regarding the educational value of social 'presence' in the form that virtual reality enables and even whether and how it might benefit institutions of higher education and their students [10] [11]. However, within a wider context researchers have demonstrated that computer-mediated communication (CMC) and multi-user virtual environments (MUVE) are capable of projecting social presence. It has also been argued that MUVEs offer more presence affordances than other forms of CMC in that they are designed to foster social interaction and the formation of groups and communities. They have the potential to "significantly reduce the subjective feelings of psychological and social distance, often experienced by distance education participants" [10]. Thus, offering courses via SL would allow for a rich and compelling learning environment while maximizing distance learning benefits, such as reaching nontraditional students and promoting international collaborations.

The 'distance' in distance education implies that physical and geographical separation is correlated with psychological and social distance. It is therefore tempting to assume that students feel disconnected and isolated from the instructor as the physical distance grows between them. However, the nature of the technology or medium used in delivering instruction possesses its own distance measure

[12] and Moore suggests it being more useful to consider distance education as pedagogical distance [13]. Moore also argues that pedagogical, or 'transactional distance' (TD) is a function of two sets of variables, *structure* and *dialogue* ('constructive interaction'). Hence, the manner in which a program is designed and conducted can result in higher or lower levels of dialog between the learner and the instructor.

3.1 Immediacy

While TD refers to pedagogical distance, and is dependent on three dimensions—structure of the program, dialogue between teacher and learner, and social presence—immediacy focuses more on the dialogue part of TD. Immediacy is the perception of physical or psychological closeness between communicators and is observed by approach and avoidance mannerisms which include verbal and non-verbal behaviors [14]. Within this framework, immediacy is a set of measures of behaviors employed in association with instructional transactions. Research on instructor immediacy suggests strongly that teachers adopting appropriate immediacy behaviors facilitate interaction and reduce psychological distance [15]. New interactive and immersive technology such as SL may enable more immediate instructional transactions between teacher and learners than traditional online platforms. Immediacy is a variable of social presence, a construct that is also influenced by the amount of information transmitted, words conveyed, and the context of the communication.

3.2 Social Presence

In some ways, the rise of virtual realities and allied new media reopen debates of the 1980s and 90's between Richard Clark, Robert Kosma and others [16] in which advocates Clark's position generally claimed that media functions primarily as conduits for instructional strategies and had few instructional effects of themselves. Kosma and his supporters [12] argued that different media enabled different and often specific instructional strategies and that some media were more effective enablers of some strategies. More importantly, Kosma believed that emerging digital multimedia would be able to approximate or stimulate many media modalities (e.g., audio, video, text, print, photos, video). These arguments foreshadowed current debates about what a 3D persistent virtual world adds to the teaching and learning experience. We are now questioning how to achieve quality and effectiveness of presence in education when mediated in SL. Arguably, the "immersiveness" of SL would constitute a psychological advantage.

Social presence reflects the degree to which a person is perceived as 'real' in a mediated communication. Social presence theory is a seminal theory of the social effects of communication technology [17]. Social presence is conceived to be a subjective quality of a medium that cannot be defined objectively. Short et al. [23] regard social presence as a single dimension that represents a cognitive synthesis of several factors such as capacity to transmit information about facial expression, direction of looking, posture and non-verbal cues as they are perceived by the individual to be present in the medium. These factors affect the level of presence that is the extent to which a medium is perceived as sociable, warm, sensitive, personal or intimate when it is used to interact with other people. Virtual reality (VR) technology is about 'being there': presence is therefore partly to do with the technology and partly to do with the users' state of mind.

3.3 Co-presence

Social presence is the feeling that other persons are present even though the characteristics and behaviors of those persons may be represented and observed via mediated communication rather than physical proximity and direct observation. Schroeder [22] suggests that more immersive VR systems enable a greater sense of presence and co-presence. However, the technology of the virtual environment can influence what the participant does: "the person using the desktop system [such as Second Life] may focus on communication, whereas the more immersed person may focus on navigating and manipulating the objects". Technological effects also exist within lower-end systems such as internet-based desktop virtual worlds: bandwidth, communication capabilities, and ease of navigation. Consequently, certain technology, social factors and personal skills might interfere with the creation and maintenance of interpersonal relationships and reduce co-presence. Schroeder [21] also identifies differences in co-presence variables based on short-term interaction or long-term interaction. Research on short-term interaction might investigate common foci of attention, mutual awareness and collaborative task performance whereas research on long-term interactions might investigate phenomena such as persistence of character, of groups, and of the environments; choice of social rules and conventions; and the relation between real and virtual.

3.4 Instructor Immediacy

Although co-presence is essential to the creation of a sense of classroom communities or learning communities, the role of the teacher or the instructor (as a co-present agent) in virtual learning environments is not well-researched. Mehrabian [14] introduced the concept of immediacy as an indicator of attitudes in verbal communication. He defines immediacy as the measure of the psychological distance which a communicator puts between himself and the object of his communication [19]. He also refined the concept of immediacy in terms of 'principles of immediacy,' which states that "people are drawn toward persons and things they like, evaluate highly, and prefer; and they avoid or move away from things they dislike, evaluate negatively, or do not prefer" [20]. Just as instructor behaviors or lack thereof may influence physical approach and avoidance behaviors, they can also be conceived as an influence on the psychological distance between people [15][20]. Thus, immediacy can be thought somewhat metaphorically as the perception of physical and psychological closeness between communicators. Verbal immediacy behaviors include calling students by name, using inclusive pronouns (e.g., 'we' rather than 'I'), inviting the use of one's first name, participating in unrelated small talk, using humor, providing feedback to students, and asking students for feedback. Nonverbal immediacy behaviors include gestures, vocal variety, smiling at students, displaying a relaxed body posture, moving around the classroom, speaking with outline only, removal of barriers, appropriate touch and professional casual dress [15] [21] [22].

3.5 Presence in Second Life

Second Life provides similar audio presence to video conference style online learning but differs in that there is no video of the participants, rather people are replaced by their graphical animations; avatars. The simulated physicality of virtual worlds and the embodied presence of avatars as agents of users facilitate behavioral displays and the appropriate adjustment of these displays to psychological circumstances in real time. This enables user expression via the avatar of behaviors communicating internal states. The avatar may also display behaviors (as an actor would) that are appropriate to a situation, but are acted or faked. User vocal expressions can be projected almost unaltered into Second Life and appear to other observers to be collocated with the user's avatar. Body language and facial expressions are either expressed autonomously by the avatar's software routines (e.g., low-level gesturing with hands, blinking and slight smiling), Eyes generally gaze in a direction determined by cursor location, reflecting mouse position. More explicit facial displays and body movements such as laughing or frowning, hand waving, or pointing require explicit execution by the user of keyboard short cuts. Thus, with current SL technology, the appropriateness of avatar expression is to a considerable degree a practiced keyboard skill rather than a direct projection of bodily movements. One implication of this current state of the art, is that instructors might exaggerate expressions, or alternatively elect expressions that do not reflect their current 'true' dispositions. In any case, instructors skilled in SL technique are well equipped to control the display of immediacy behaviors of their avatars and thus potentially control the psychological distance between them and the students.

4 Conclusion

In this paper we have explained how there is an increasing interest in extending the use of technically enabled environments commonly found in AmI, from physical to the virtual. We have introduced tools such as SecondLife and its derivative, RealXtend that enables virtual worlds to be built. We also explained how regular physical AmI environments can be linked to their virtual counterparts creating so-called mixed reality environments. We described an important application of such environments, distance learning, and discussed the issues that influence its effectiveness; notably social presence and immediacy, explaining how technology may enhance or detract from these. In particular we noted that whereas most research questions of the last century regarding educational implications of immediacy and social presence focused on the instructor as the person of interest, social networking software that connects hundreds of millions of users online demands the traditional focus be broadened to all members of learning communities. From our work it is evident that virtual reality-based avatars challenge early paradigms for research on social presence and immediacy in two ways: the source of communication control and the dominant instructor as source of immediacy. The Networked Minds paradigm exemplifies new lines of inquiry that emerged in the 1990's that extend beyond immediacy behaviors to measure emotional and cognitive states, and collaborative dispositions. With these new perspectives and new instrumentation, researchers will better be prepared to investigate complex communication modalities and media that integrate and filter sensorimotor, cognitive, and affective cues of communicators which will all need to be accounted for in mixed reality AmI social spaces, such as online learning.

Clearly venturing into virtual environments will expose many new challenges for AmI systems and whilst this work is at an early stage, we hope the insight provided in this paper will be helpful to those considering working in this area of AmIsystems.

References

1. Shen, R.M., Wang, M.J., Gao, W.P., Novak, D., Tang, L.: Mobile Learning in a large blended computer science classroom: System function, pedagogies, and their impact on learning. IEEE Transactions on Education 52(4), 538–546 (2009)
2. Gartner, Inc. Press Release (April 24, 2007), `http://www.gartner.com/it/page.jsp?id=503861` (retrieved December 14, 2007)
3. Lenhart, A., Kahne, J., Middaugh, E., Macgill, A., Evans, C, Vitak, J.: Teens, video games, and civics. In: Pew Internet and American Life Project, Washington, DC (2008), `http://www.pewinternet.org/pdfs/PIP_Teens_Games_and_Civics_Report_FINAL.pdf` (retrieved September 19, 2008)
4. Davies, M., Callaghan, V., Gardner, M.: Towards a mixed reality intelligent campus. In: Proceedings of the 2008 International Conference on Intelligent Environments (IE 2008), Seattle, USA (2008)
5. Meadows, M.S.: I, Avatar: The culture and consequences of having a second life. New Riders, Berkeley (2008)
6. RealXtend, realXtend – Open source platform for interconnected virtual worlds, `http://www.realxtend.org` (retrieved March 1, 2010)
7. OpenSim, Main Page – OpenSim, `http://opensimulator.org` (retrieved February 22, 2010)
8. Linden Lab., Second Life, `http://secondlife.com/` (retrieved February 15, 2010)
9. Google, 3D Warehouse, `http://sketchup.google.com/3dwarehouse` (retrieved February 22, 2010)

10. Callaghan, V., Gardner, M., Horan, B., Scott, J., Shen, L., Wang, M.J.: A mixed reality teaching and learning environment. In: Fong, J., Kwan, R., Wang, F.L. (eds.) ICHL 2008. LNCS, vol. 5169, pp. 54–65. Springer, Heidelberg (2008)
11. Bower, J., Christensen, C.: Disruptive Technologies: Catching the Wave. Harvard Business Review (March 3, 2009) (HBR OnPoint Enhanced Edition), http://www.amazon.com/Disruptive-Technologies-Catching-OnPoint-Enhanced/dp/B00005REGO (retrieved February 8, 2010)
12. Kosma, R.: Will media influence learning? Reframing the debate. Educational Technology, Research & Development 42(2), 7–19 (1994)
13. Moore, M.: Theory of transaction distance. In: Keegan, D. (ed.) Theoretical principles of distance education, pp. 22–29. Routledge, New York (1993)
14. Mehrabian, A.: Immediacy: An indicator of attitudes in linguistic communication. Journal of Personality 34, 26–34 (1966)
15. Andersen, J.: Teacher immediacy as a predictor of teaching effectiveness. In: Nimmo, D. (ed.) Communication yearbook, vol. 3, pp. 543–559. Transaction Books, New Brunswick (1979)
16. Hastings, N., Tracey, M.: Does Media Affect Learning: Where Are We Now? TechTrends 49(2), 28–30 (2004)
17. Short, J., Williams, E., Christie, B.: The social psychology of telecommunications. John Wiley & Sons, London (1976)
18. Schroeder, R.: The social life of avatars. Springer, London (2002)
19. Mehrabian, A.: Inference of attitudes from the posture, orientation, and distance of a communicator. Journal of Consulting and Clinical Psychology 32, 296–308 (1968)
20. Mehrabian, A.: Silent messages. Wadsworth, Belmont (1971)

A Digital Secretary for Smart Offices Setup Up

João Laranjeira, Carlos Freitas, Goreti Marreiros,
Carlos Ramos, and João Carneiro

Abstract. In Ambient Intelligence paradigm, helping people in their daily routine tasks is a priority that has to be accomplished. People have less availability and the number of devices is growing, so some kind of autonomy must be given to the environment. In this paper we expose a system able to prepare meeting rooms for the execution of several kinds of events. Our proposal is based on an agents' community that performs in an autonomous way the operations required to setup the meeting room to a specific, including the configuration of software and hardware. The agents are autonomous and fulfilled with capabilities to prevent software and hardware failing. The proposed digital secretary was tested in LAID (Laboratory of Ambient Intelligence for Decision making) present in GECAD and it was able to promote the environment autonomy.

1 Introduction

Ambient Intelligence (AmI) deals with a new world where computing devices are spread everywhere (ubiquity), allowing human being to interact with physical world environments in an intelligent and unobtrusive way [9]. AmI can be introduced in several different environments and involves many different disciplines, like automation (sensors, control and actuators), human-machine interaction and computer graphics, communication, ubiquitous computing, embedded systems and, obviously, Artificial Intelligence. In the aim of Artificial Intelligence, research expects to include more intelligence in the AmI environments, allowing a better support to the human being and the access to the essential knowledge in order to make better decisions when interacting with these environments [9].

Carlos Freitas · Goreti Marreiros · Carlos Ramos
Institute of Engineering – Polytechnic of Porto

João Laranjeira · Carlos Freitas · Goreti Marreiros · Carlos Ramos · João Carneiro
GECAD – Knowledge Engineering and Decision Support Group
e-mail: `jpcl@isep.ipp.pt`, `cff@isep.ipp.pt`,
`goreti@dei.isep.ipp.pt`, `csr@dei.isep.ipp.pt`

P. Novais et al. (Eds.): Ambient Intelligence - Software and Applications, AISC 92, pp. 69–76.
springerlink.com

Group decision making is, by definition, an excellent area that demonstrates the potential of Smart Meeting Rooms (SMR) [6]. Considering group decision making, in which meetings will be distributed by various places and/or in rooms filled with devices and sensors, meetings places arrangement tend to be monotonous and time consuming. Smart Offices (SO) concepts, a subdiscipline of AmI, emerged with the promise to optimize/improve interactions with such tools and also to minimize the users' effort [6]. We will now analyse some of the definitions presented in literature. Le Gal C [5] defined a Smart Office as an environment that is able to help its inhabitants to perform everyday tasks by automating some of them and making the communication between user and machine simpler and effective. Marsa-Maestre et al [7] defined Smart Office concept as an environment that is able to adapt itself to the user needs, to release the users from routine tasks they should perform, to change the environment in order to suit their preferences and to access services available at each moment by customised interfaces. Ramos et al [10] defined Smart Office concept as an environment that is able to reduce the decision-cycle offering, for instance, connectivity wherever the user is, aggregating the knowledge and information sources.

There are several projects exploring SO concept and Active Badge [11] was the first approach. Monica Project [5] intends to anticipate user intentions and augment the environment to communicate useful information through user monitoring. In Intelligent Environment Laboratory of IGD Rostock [3] is intended to create an interactive environment based on multimodal interfaces and goal-oriented interaction differing from other systems that normally use a function-oriented interaction. Sensor-R-Us is placed in the University of Stuttgart [8] and this application is useful in order to know the position and status of persons and other information about the room, such as the temperature, its availability or the number of meetings that a person has in a specific day. EasyMeeting [1] explores the use of FIPA agent technologies, Semantic Web ontologies, logic reasoning, and security and privacy policies. Its goal is to create a smart meeting room that can facilitate typical user activities in an everyday meeting. Considering that routine tasks must be performed in an autonomous way we present an agent based on digital secretary that aims to prepare a SMR for the reception of events, for instance presentations, decision-making meetings using GDSS's. Such events require tasks like connecting servers and workstations, running applications that are used in process, connecting/adjusting audio and cameras. We aim to set up such routines in an autonomous way. With this digital secretary the SMR will automatically adapt itself to an event context. The user can perform such action by scheduling events or triggering them when he is inside of the room. A scalable approach, give to users the power to build new routines and increase the room autonomy were our main goals. In this paper we will demonstrate the system architecture and the users' interfaces. Then, in experiments section, the digital secretary that we propose in this paper is applied and tested at LAID, a SMR presented in GECAD. Conclusions and future work are given in the last section.

2 MeetingRoomSetup System

MeetingRoomSetup System aims to give meeting rooms the ability to prepare itself for the reception of events. Implemented events include presentations, meeting's assisted by GDSS, idea generation meeting and choose and selection of alternatives meeting. All mechanisms will be activated to perform the tasks required to trigger the event. These processes are performed by the agents' community that approach the problem in a cooperative way, in order to adapt the environment to the desired context in an autonomous way.

In order to achieve these goals we propose a model based in 3 concepts: *Tasks, Routine and Services*. *Tasks* represent all the possible events that can occur, such as presentations, idea generation meetings, choose and selection of alternatives meeting, etc. *Routines* can be seen as repetitive operations that must be performed, launching a software, hardware or alerts, run the same applications in several computers, connect some computers, send alerts to the administrator. Finally, *services* are atomic operations which have as a function the activation of software, hardware or sending an alert, such as, connecting this computer, running a specific application in a specific computer or sending email to administrator.

In terms of architecture MeetingRoomSetup is composed by two main components: MeetingRoomSetupWeb and a community of agents. There are three secondary components: a database, an application of management for agents and an application of BackOffice to the management of the system data.

In the following subsection we explore the proposed architecture, its components and the interactions among different components.

2.1 Architecture

Fig. 1 represents the architecture of the system. The architecture is composed by an agents' community implemented using Open Agent Architecture (OOA) framework, a relational database and several interfaces of access to different users.

In the agents community we have introduced 3 kinds of agents, including task agents that represent events that can occur in meeting rooms; routine agents who represent a group of tasks that need to be performed in order to achieve a task, which can connect computers, servers, run applications, etc; and service agents that represent small services, for example connect a computer, send email to system administrator, turn the sound on, etc. Such agents' structure enables the users to define new events that the environment can receive; the administrator can perform such task by reusing routines or even by personalizing new ones by selecting small operations (services) already existent. They are also able to develop new agents and introduce them in the environment and in the community. The agents' community is coordinated by the agent facilitator (1), responsible to deliver the messages to the right agent. Computers present in the environment are activated by agents inside the community, once started they have agents who run locally (agents service). The purpose of these agents is to execute operations that must be performed locally.

Agents' community interaction with the user is performed by the following modules: MeetingRoomSetupWeb, AgentsManagerApplication and OAAMonitor. The first one is in a webpage format. The second one is an application that aims to register agents in community of agents and to help the system administrator to test tasks activation. OAA Monitor application is an interface that allows monitoring community of agents.

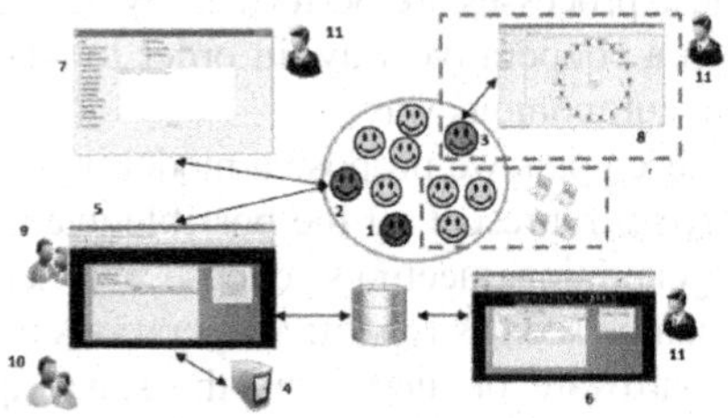

Fig. 1 Digital Secretary Architecture

All data circulating in the system is represented by MeetingRoomSetupDB component. It's a database which stores all data of the system. This data will be operated by some of the agents, web interface and SmartBackOffice. Users are directly involved with the system. They are divided in 3 groups: intern users, external users and administrator user. The first and second type have access to interface web with the aim to consult some data of the system, but the main goal is to benefit from the main functionality of system (tasks scheduling and tasks execution). The role of the administrator is to manage the community of agents and manipulate the data in the system. For that he will use the AgentsManagerApplication (community of agents' management) and SmartBackOffice (data manipulating).

2.2 Community of Agents

As we have seen in the previous section, the community of agents includes different types of agents: agent app (2), agent monitor (3), agent facilitator (1), agent task, agent routine and agent service. Agents' community was developed with the OAA framework and agents were implemented in JAVA. Agent App is an interface agent. It makes a bridge of communication between the agents' community and the interface component and vice versa. Interface components are AgentsManagerApplication and MeetingRoomSetupWeb. Agent Monitor is responsible for the control and management of all agents presented in the community of agents. This application will help the administrator of the system to control and manage the community of agents. Agent Facilitator is coordinating the community of agents. Each agent of the community of agents registers their abilities or features in the Agent Facilitator. When services are requested by an agent, instead of asking to a specific agent that performs this task, the agent simply make the request to the Agent Facilitator and this one decides which agents are available to that task. Agent task represents a task that the system can solve. At this moment the digital secretary is able to perform 3 different tasks: presentation, idea generation and idea selection. For each task there is an agent task associated. They communicate with others agents with the purpose of these agents activate the system. Agents task knows the start date, end date and all resources needed to start the task. Agent Routine represents a routine in the system. The examples of routines are: connect computers, send alerts, run applications, among others. Each agent routine is responsible for one routine. For example, "agent alert" knows that we must alert 5 users, so he send 5 requests to "agent email" that is responsible by the service of sending emails. Agent Service represents a service in the system. Examples of services are: connect computer 12, run Firefox browser, send email to user admin, among others. Agent service receives a request of agent routine and executes this request with the purpose of preparing the room for an event. These agents are terminal agents because they don't send request to others agents - they execute services. There is an agent service that is special – the AgentSearchTasks. This agent constantly searches for scheduled tasks. When this agent finds a task, launches it in system then. All agents of this community are prepared to prevent software and hardware's failure that affect the proper operation of an event. For example, if a computer fails then the agent will connect other computer to prevent a wrong task operation. The main motivation to use agents in this system was to take advantage of the characteristics of persistence, autonomy and proactivity associated with the agents. Other components allow maintaining a scalable system that features can be extended by users. Together they endow the environment with the capability to resolve problems that may arise.

2.3 MeetingRoomSetupWeb

The aim of MeetingRoomSetupWeb (5) is to help the user to get all functionalities of the system. Users (9) inside the environment are able to access all functionalities of the system, whereas the external user (10) may only access scheduling functionalities, view tasks and cancel tasks. Through this application, administrators are able to control all tasks carried. External users must pass through an LDAP server (4) authentication. Beyond functionalities of external user, internal user may start or stop a specific task. Internal users are considered authenticated just by being in the environment and have access to MeetingRoomSetupWeb.

In Fig. 2 we can see the connection of the computer's process. In this example the problem that we want resolve is to prepare our test case environment, LAID, to a presentation. Before this process, a task agent (AgentPresentation) receives a request from the system to start a presentation. In the first step of this process, AgentConnectPCs (agent routine) requests to local agent (AgentConnectPC) to connect and this agent "wakes-up" the computer (PC22). After this, AgentConnectPCs (agent routine) verifies if the local agent (AgentConnectPC) is active. If the agent is unavailable then the AgentConnectPCs sends the request to another AgentConnectPC (agent service) that is available. This process is repeated until it finds a replacement computer or until it finishes the available resources.

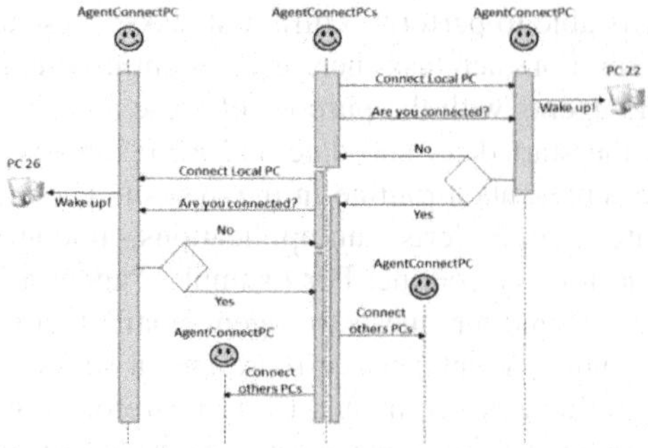

Fig. 2 Connection of computer process

2.4 SmartBackOffice and AgentsManagerApplication

SmartBackOffice application (6) allows the administrator to manipulate the data base of the system. It allows setting up several events, creating new ones by using the routines and services existent. It also allows adding of new services that included the addition of new agents with new kinds of services. This application allows scheduling an event to a certain date, so the system will be automatically activated on the chosen date. AgentsManagerApplication (7) is a very simple application that allows the administrator to run all agents required to execute all the tasks available. So this application will have a very simple interface where there are multiple options, for example, to create or delete agents. OAA Monitor (8) is an application that helps AgentsManagerApplication. OAA Monitor has the role of showing in a graphical way, all registered agents in the community of agents. The administrator can create new agents to add new tasks, routines and services to the system. For example, if a new device was added to the environment, the administrator could create a new agent to control the device.

3 Experiments

To evaluate this system, we are going to present a scenario and a problem that will be resolved. The goal is to prepare the LAID to an idea generation meeting. During this experiment, agent task will delegate procedures to agents routines that will execute them. The agent connectPCs (agent routine) will control the connection of computers. If any computer doesn't connect then this agent will contact and connect another one. To complete this task it is necessary to fulfil the following routines and services:

- Idea generation Meeting (Task)
  - o Connect LAID's computers for 4 decision points (Routine);
    - ▪ Connect PC21, PC22, PC23 and PC25 (4 Services);
  - o Run needed applications (Routine);
    - ▪ Run IGTAI [2] on all computers (4 Services);
  - o Connect sound and video (Routine);
    - ▪ Connect all cameras and audio devices (2 Services);

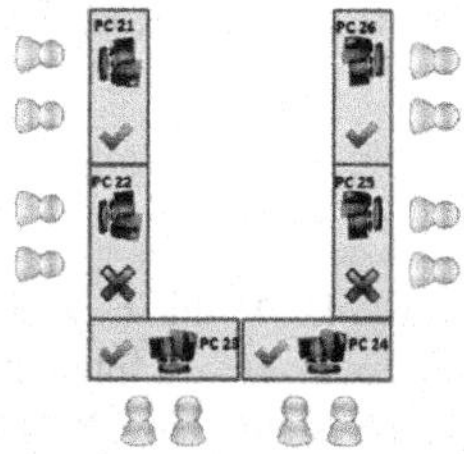

Fig. 3 Process of connecting LAID's computers

In Fig. 3 we can see the computers and users distribution in LAID. In this figure is represented the process of computers connection. PC22 and PC 25 are unavailable. AgentConnectPCs will detect this problem and connect two available computers, for example, PC 24 and PC26. Therefore the system adapts to the raised problems. To complete the task, AgentRunApplications will run applications on these computers. To evaluate this system, we compare it with iTalc [4] (Intelligently Teaching And Learning with Compute) and with a manual preparation of LAID. In Table 1 we can see the time needed to prepare the LAID for the scenario presented above. The results obtained with the MeetingRoomSetup System were considerably better compared to the other two options analysed.

Table 1 Time needed to prepare the LAID

	Time needed to prepare the LAID (min)
MeetingRoomSetup System	4
iTalc [4]	13
Manual Preparation	27

4 Conclusions and Future Work

In this paper was proposed a digital secretary endowed able to prepare a meeting room to receive different kinds of events. Realising the users from this hardware and software setup will contribute to the achievement of AmI characteristics in meeting rooms. The use of cooperative agents in the proposed architecture allows endowing the proposal with persistence, autonomy and proactivity. Giving hierarchy to agents and the ability to deploy them in different locations considering their goals allows us to build a scalable system that features can be extended by users.

As future work we want to develop the usage of semantic technologies to represent the state of the environment in real time, by forcing the agents to update a semantic representation of the environment state when they change it. We believe it will be possible to optimize agent's operations. Also, we intend to represent the users by means of agents in order to recognize automatically user's intentions, preferences and profile when they are in the meeting room.

Acknowledgments. The authors would like to acknowledge FCT, FEDER, POCTI, POSI, POCI, POSC, and COMPETE for their support to R&D Projects and GECAD.

References

[1] Chen, H., Perich, F., Chakraborty, D., Finin, T., Joshi, A.: Intelligent agents meet semantic web in a smart meeting room. In: 3rd International Joint Conference on Autonomous Agents and Multi Agent Systems, New York, USA, pp. 854–861 (July 2005)

[2] Freitas, C., Marreiros, G., Ramos, C.: IGTAI- An Idea Generation Tool for Ambient Intelligence. In: 3rdIET International Conference on Intelligent Environments, Ulm, Germany (2007)

[3] Heider, T., Kirste, T.: Intelligent environment lab. Computer Graphics Topics 15(2), 8–10 (2002)

[4] iTalc - (2011), http://italc.sourceforge.net/

[5] Le Gal, C.: Smart Environments: Technology, Protocols and Applications. Wiley, chap Smart Offices (2005)

[6] Marreiros, G., Santos, R., Ramos, C., Neves, J., Novais, P., Machado, J., Bulas-Cruz, J.: Ambient Intelligence in Emotion Based Ubiquitous Decision Making. In: Proc. Artificial Intelligence Techniques for Ambient Intelligence, IJCAI 2007 – Twentieth International Joint Conference on Artificial Intelligence, Hyderabad, India (2007)

[7] Marsa-Maestre, I., Lopez-Carmona, M.A., Velasco, J.R., Navarro, A.: Mobile agents for service personalization in smart environments. Journal of Networks (JNW) 3(5), 30–41 (2008)

[8] Minder, D., Marrón, P.J., Lachenmann, A., Rothermel, K.: Experimental construction of a meeting model for smart office environments. In: Proceedings of the First Workshop on Real-World Wireless Sensor Networks (2005)

[9] Ramos, C.: Ambient Intelligence Environments", Encyclopedia of Aritificial Intelligence. In: Rabunãl, J., Dorado, J., Sierra, A. (eds.) Information Science Reference (2009)

[10] Ramos, C., Marreiros, G., Santos, R., Freitas, C.: SmartOffices and Intelligent Decision Rooms. In: Handbook of Ambient Intelligence and Smart Environments (AISE), pp. 851–880. Springer, Heidelberg (2009)

[11] Want, R., Hopper, A., Falcao, V., Gibbons, J.: The active badge location system. ACM Transactions Information Systems 10(1), 91–102 (1992)

Applying HoCCAC to Plan Task the COPD Patient: A Case Study

Juan A. Fraile, Sara Rodríguez, Juan F. de Paz, and Belén Pérez-Lancho

Abstract. This paper presents a multiagent system that facilitates the performance of daily tasks for Chronic Obstructive Pulmonary Disease (COPD) patient within a context-aware environment. The paper analyzes the relevant aspects of context-aware computing and presents a prototype that can be applied to monitor COPD patient at their homes. The system includes computational elements that are integrated within a domestic environment with the goal of capturing context-related information and managing the events carried out by the patient. The services are support by the processing and reasoning out of the data received by the agents in order to offer proactive solutions to the user. The results obtained with this prototype are presented in this paper.

Keywords: Context-Aware, Multi-Agent Systems, Home Care.

1 Introduction

The preferred characteristics when designing software applications include autonomy, security, flexibility and adaptability. In order to achieve this objective, it is necessary to have mechanisms, methods and tools that can develop systems capable of adapting to changes within the environment. The search for flexible software applications that can continually improve their ability to adapt to the demands of the users and their surrounding leads us to context-aware systems that store and analyze all of the relevant information that surrounds and forms a part of the user environment. Context-aware systems provide mechanisms for developing

Juan A. Fraile
Pontifical University of Salamanca, c/ Compañía 5, 37002 Salamanca, Spain
e-mail: jafraileni@upsa.es

Sara Rodríguez · Juan F. de Paz · Belén Pérez-Lancho
Departamento de Informática y Automática, University of Salamanca,
Plaza de la Merced s/n, 37008 Salamanca, Spain
e-mail: {srg,fcofds,lancho}@usal.es

P. Novais et al. (Eds.): Ambient Intelligence - Software and Applications, AISC 92, pp. 77–84.
springerlink.com

applications that understand their context and are capable of adapting to possible changes. A context-aware application uses the context of its surroundings to modify its performance and better satisfy the needs of the user within that environment. The current trend for displaying information to the system users, given the large number of small and portable devices, is the distribution of resources through a heterogeneous system of information networks. Web applications and services have been shown to be quite efficient [12] in processing information within this type of distributed system. Web applications are run in distributed environments and each part that makes up the program can be located in a different machine. Some of the web technologies that have had an important role over the last few years are multiagent systems and SOA (Service Oriented Architecture) architectures, which focus on the distribution of system service functionalities. This model provides a flexible distribution of resources and facilitates the inclusion of new functionalities within changing environments. In this respect, the multiagent systems have also already demonstrated their aptitude in dynamic changing environments [2] [6]. The advanced state of development for multiagent systems is making it necessary to develop new solutions for context-aware systems. It involves advanced systems that can be implemented within different contexts to improve the quality of life of its users. There have been recent studies on the use of multiagent systems [2] as monitoring systems in the medical care [1] patients who are sick or suffer from Alzheimer's [6]. These systems provide continual support in the daily lives of these individuals [5], predict potentially dangerous situations, and manage physical and cognitive support to the dependent person [3].

This paper presents the Home Care Context-Aware Computing [9] (HoCCAC) multi agent system that supervises and monitors dependent persons in their homes, providing the user with a certain degree of self-sufficiency. The proposed system focuses on incorporating mechanisms that facilitate the integration of web applications. The HoCCAC system provides wireless communication between its elements, and integrates intelligent agents with sensors and autonomous components that obtain context-aware information and are proactive in their interaction with the users. HoCCAC facilitates the automation of context devices and the ability to respond to the elements from a remote location. One of the most important characteristics of HoCCAC is the use of intelligent agents as one of the principal components.

The remainder of the paper is structured as follows: section 2 presents the agents in HoCCAC system. Section 3 describes a study case where the HoCCAC system is applied. Finally, section 4 presents the results and conclusions obtained after evaluating the study case.

2 Agents in HoCCAC

HoCCAC can be defined by the need to control various devices and gather user information in a non-intrusive and automatic way within Context-Aware environments [9]. HoCCAC makes it possible to automatically obtain information on users, their actions and environment in a distributed manner. HCCAC primarily focuses on monitoring a person in their home and sending notifications on the state of the individual or possible incidents. Additionally, it combines the management of personal information with a model of daily activities that are developed by using the data provided by the sensors through the household network. The interpretation and reasoning of the base knowledge and the daily activity models developed by the system provide an added value to the system. HCCAC makes is possible to easily use and share context-aware applications within changing physical spaces. As seen in Figure 1, the system is composed of the following agents:

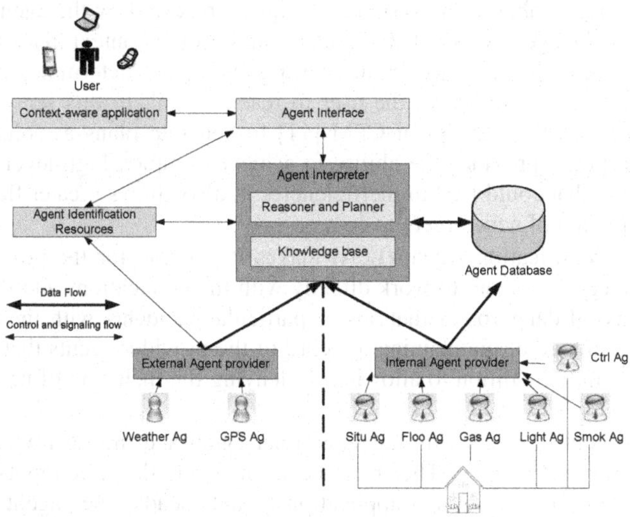

Fig. 1 Overview of the multi-agent system HoCCAC

- Provider agents capture and summarize the data obtained by both internal and external heterogeneous context sources so that the Interpreter and Database agents can process and reuse these data. The system is dynamic and is capable of incorporating an information Provider agent at any given moment by adding the corresponding sensor or capturing the necessary information from a server or external provider. The types of provider agents can connect to the system are: (i) agent situation, which continuously maintains the patient located at home, (ii) agent detector flood, (iii) gas detector agent, (iv) agent Smoke detector, (v) light sensor agent, (vi) agent thermostat, (vii) control agent, which establishes and controls the daily parameters desired by the patient, (viii) weather agent, which records the time abroad to act accordingly on the internal variables at home and (ix) GPS agent, which maintains the patient located outside the home. The system is dynamic and has the ability to incorporate information provider agent at any time by adding the appropriate sensor or capturing the necessary information from a server or external supplier.

- Database agents store the context data obtained by the Provider agents. The organization of this information is similar for different environments. This agent is in charge of managing and exploiting the stored data. Additionally it provides the necessary information to the interpreter agent and records and updates the action plans that are determined by the interpreter agent.
- Interpreter agent provides logical reasoning services to process the contextual information. The interpreter agent reasons out the actions that must be taken by creating a plan that determines the optimal course of action for reaching an objective. This agent provides the ability to reach high level and complex objectives and avoids errors that could result in inefficacies. It also allows for greater flexibility when dealing with new objectives. The interpreter agent, using information available in the system, the signals received by the agent interface and the knowledge base and beliefs regarding keeping context, reasoning about actions to take. In this way creates an argument that determines an optimal course of action or plans for the user to reach a goal. In this sense the agent interpreter uses the concept of CPM [11] to generate plans as solutions. The agent interpreter provides the ability to achieve complex high-level goals and avoid errors that could lead to inefficiencies. It also allows greater flexibility in the presentation of new targets.
- Resource Identification Agent (LcA) makes it possible for the provider agent and the interpreter agent to work directly with the applications and the users in order to avoid dangerous situations or particular incidents with the user. This agent is in charge of maintaining a record of the provider agents that are active in the system, in addition to allowing or denying the inclusion of new provider agents.
- Interface agent interacts with the Interpreter agent and the LcA agent without explicit user instructions. The interface agent reads the user inputs provided through the context-aware applications, and sends the agent behavior modification interpreter agent or LcA agent. The interface agent also sends notifications to users through context-aware applications. The agent interface for example, can receive many entries context-aware applications over a long period of time, before deciding to take a single action, or entry of a single context-aware application can launch a series of actions by the agent.

Context-aware applications in HoCCAC also check the information available from the context providers and are in constant listening mode to deal with possible events that the context providers transmit. The applications use different levels of context information and adapt their behavior according to the active context. They check the functionalities registered in the system and have a location for the entire context providers made available within the environment.

The agents described are independent of the platform on which they are installed. The external provider agents obtain context information through external resources such as, for example, a server that provides meteorological information about the weather in a specific place, or a location server that provides information on the location of a person who is not at home. The internal provider agents gather information directly from the sensors installed within the environment, such as RFID based location sensors installed in the home of a patient, or light sensors.

The functions of the interpreter agent include both processing information provided by the database agent, and reasoning out the information that has been processed.

3 Case Study: Applying HoCCAC to Plan Task the COPD Patient

A prototype has been developed to improve the patient quality of life at home using HoCCAC system. The system collects information from sensors and interacts with the patient through plans of tasks. The information collected by the sensors is primarily the user's location-aware in the environment. In addition the system also collects information about the temperature on the rooms of the patient's home and the state of the lights in areas for which the user travels. The prototype obtains information of the environment and plans tasks for COPD patients, trying to improve the living conditions of the patient.

The interpreter agent receives a set of key activities to be performed by a COPD patient at home. This list of activities is responsible for defining the medical staff monitors the patient's condition. The table 1 shows the list of core activities for a COPD patient. With the activities list defined and the medical personnel, the interpreter agent generates a plan following the CPM method. Prior activities list to each activity and the times list can be determined by the duration of the task execution plan and develop the work plan's network. Table 1 shows the activities list and their predecessors.

Table 1 Activities list for COPD patients and its predecessors

Activity	Predecessors	Name	Time (min.)	Task
A	-	Oxygen cylinder	600	1
B	-	Wake up and exercise	10	6
C	-	Breakfast	10	10
D	A	Pill and spray at 8 hour	3	3
E	A, B, C, D	Doctor Visit	60	14
F	E	Walk	30	4
G	F	Lunch	20	9
H	G	Oxygen cylinder	300	2
I	G	Pill and spray at 14 hours	3	3
J	I	Lunch	40	10
K	J	Picnic	20	11
L	H, K	Walk	30	5
M	L	Pill and spray at 20 hours	3	3
N	M	Dinner	30	12

Once the interpreter agent has defined the work plan and the duration of the plan, it transfers this information to the rest of the agents in the HoCCAC system to execute the plan. The provider agents together with the interface agent are responsible for monitoring compliance with the plan of work. If at any time there was an interruption of the plan, for example, because the patient has a choke and has to do the activities specified for this case the system receives notice HoCCAC through the interface agent and immediately interpreter agent makes a redevelopment plan task. Another feature is given by the doctor visits, as are monthly. Depending on the type of interruption, the provider agent can also generate a notice, for example, in the case of failure of the oxygen tank.

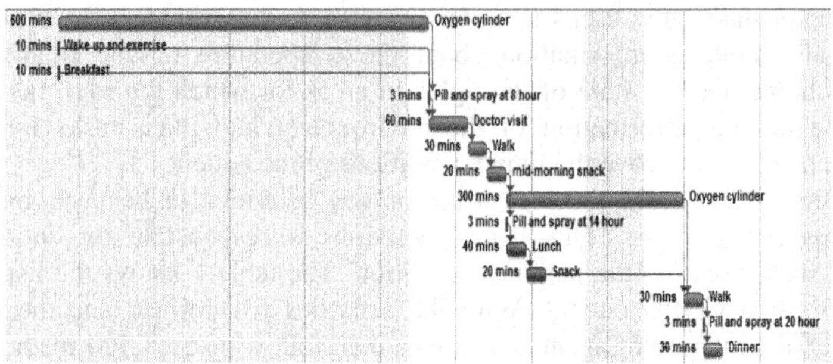

Fig. 2 Gantt chart associated with task plan

Figure 2 shows the Gantt chart with the proposed plan and the associated tasks. In addition Figure 2 also shows the activities that are critical and which not. In red are the critic task and blue are the task with slack. Figure 2 also shows the overlapping of tasks. For example, getting the oxygen cylinder up, exercise and breakfast and it also follows that a doctor's visit comes after the above activities, plus that of taking the medication at 8 am. During the entire sequence of operations performed by HoCCAC agents must take into account the transformation of information that occurs in the system. The information provider agents, together with the interpreter agent are responsible for carrying out this work. Furthermore, the patient at all times can interact with the context to set the parameters that govern the functioning of HoCCAC through context-aware applications.

4 Results and Conclusions

HoCCAC was used to develop a prototype used in the home of a dependent person. It incorporates JavaCard technology to identify and control access, with an added value of RFID technology. The integration of these technologies makes the system capable of automatically sensing stimuli in the environment in execution time. As such, it is possible to customize the system performance, adjusting it to the characteristics and needs of the context for any given situation.

Different studies related to context-aware systems, such as [4] [7] [12], focus exclusively on gathering positional data on the user. The authors of these papers gather the positional data on the users through GSP signals, mobile telephone towers, proximity detectors, cameras and magnetic card readers. Many of these signals work with a very wide positioning range, which makes it difficult to determine the exact position of the user. In contrast, the system presented in this paper determines the exact position of the user with a high level of accuracy. To do so, the system uses JavaCard and RFID microchip located on the users and in the sensors that detect these microchips in their context. Others studies, such as [10], in addition to locating the users in their context, try to improve the communication between patients and medical personnel in a hospital center by capturing context attributes such as weather, the state of the patient or role of the user. In addition to capturing information from various context attributes such as location, temperature and lighting, HoCCAC also incorporates the Interpreter agent reasoning process to provide services proactively to the user within a Home Care environment. At the same time offers the patient proactive work plans that seek to improve the patients' quality of life. The user can perform their daily tasks and receive support intelligent context without explicit interaction. Therefore, the user does not need to learn to use the system. This makes the degree of user satisfaction with the system which handles HoCCAC, increases. HoCCAC incorporates new information Provider agents in execution time. In this respect, HoCCAC proposes a model that goes one step further in context-aware system design and provides characteristics that make it easily adaptable to a home care environment.

Although there still remains much work to be done, the system prototype that we have developed improves home security for dependent persons by using supervision and alert devices. It also provides additional services that react automatically in emergency situations. As a result, HoCCAC creates a context-aware system that facilitates the development of intelligent distributed systems and renders services to dependent persons in their home by automating certain supervision tasks and improving quality of life for these individuals. The use of a multi-agent system, web services, RFID technology, JavaCard and mobile devices provides a high level of interaction between care-givers and patients. Additionally, the correct use of mobile devices facilitates social interactions and knowledge transfer. Our future work will focus on obtaining a model to define the context, improving the proposed prototype when tested with different types of patients.

Acknowledgements. This study has been supported by the Spanish Ministry of Science and Innovation project TRA2009_0096.

Referentes

1. Angulo, C., Tellez, R.: Distributed Intelligence for Smart Home Appliances. On: Tendencies of data mining in Spain. In: Spanish Data Mining Network, Barcelona, Spain (2004)

2. Ardissono, L., Petrone, G., Segnan, M.: A conversational approach to the interaction with Web Services. In: Computational Intelligence, vol. 20, pp. 693–709. Blackwell Publishing, Malden (2004)
3. Bahadori, S., Cesta, A., Grisetti, G., Iocchi, L., Leone1, R., Nardi, D., Oddi, A., Pecora, F., Rasconi, R.: RoboCare: Pervasive Intelligence for the Domestic Care of the Elderly. AI*IA Magazine Special Issue (January 2003)
4. Burrell, J., Gay, G.: E-graffiti: evaluating real-world use of a context-aware system. Interacting with Computers – Special Issue on Universal Usability 14(4), 301–312 (2002)
5. Corchado, J.M., Bajo, J., de Paz, Y., Tapia, D.: Intelligent Environment for Monitoring Alzheimer Patients, Agent Technology for Health Care. Decision Support Systems 34(2), 382–396 (2008) ISSN 0167-9236
6. Corchado, J.M., Bajo, J., Abraham, A.: GERAmI: Improving the delivery of health care. IEEE Intelligent Systems. Special Issue on Ambient Intelligence (March/April 2008)
7. Espinoza, F., Persson, P., Sandin, A., Nyström, H., Cacciatore, E., Bylund, M.: GeoNotes: Social and navigational aspects of location-based information systems. In: Abowd, G.D., Brumitt, B., Shafer, S. (eds.) UbiComp 2001. LNCS, vol. 2201, pp. 2–17. Springer, Heidelberg (2001)
8. Fraile, J.A., Bajo, J., Corchado, J.M.: Multi-Agent Architecture for Dependent Environments. Providing Solutions for Home Care. Inteligencia Artificial 13(42), 36–45 (2009); Special Issue 7th Ibero-American Workshop in Multi-Agent Systems (Iberagents 2008)
9. Fraile, J.A., Bajo, J., Corchado, J.M.: Applying context-aware computing in dependent environments. In: Mira, J., Ferrández, J.M., Álvarez, J.R., de la Paz, F., Toledo, F.J. (eds.) IWINAC 2009. LNCS, vol. 5602, pp. 85–94. Springer, Heidelberg (2009)
10. Muñoz, M.A., Gonzalez, V.M., Rodríguez, M., Favela, J.: Supporting context-aware collaboration in a hospital: An ethnographic informed design. In: Favela, J., Decouchant, D. (eds.) CRIWG 2003. LNCS, vol. 2806, pp. 330–344. Springer, Heidelberg (2003)
11. Aquilano, N.J., Smith, D.E.: A formal set of algorithms for project scheduling with critical path scheduling/material requirements planning. Journal of Operations Management 1(2), 57–67 (1980)
12. Oren, E., Haller, A., Mesnage, C., Hauswirth, M., Heitmann, B., Decker, S.: A Flexible Integration Framework for Semantic Web 2.0 Applications. IEEE Software 24(5), 64–71 (2007)

Elder Care's Fall Detection System

Filipe Felisberto, Nuno Moreira, Isabel Marcelino,
Florentino Fdez-Riverola, and António Pereira

Abstract. With the increase of the elderly population, new challenges to enable a healthy and dignified life for the elderly arise. One of these challenges, comes from a very serious problem to which the elderly population is subject: falls when they are alone. This article intends to present the initial study performed, and the resulting architecture of a complete system, that in conjunction with the rest of the Elder Care project, will enable the rapid detection of falls and sending of requests for help, that may well save lives.

Keywords: fall detection, body area network, health monitoring, aging.

1 Introduction

The populations' aging tendency, which is more significant in first world countries, is also an alert to the necessity of adapting existing technologies to the physical challenges that aging entails. A recent Eurostat study predicted

Filipe Felisberto · Nuno Moreira · Isabel Marcelino · António Pereira
School of Technology and Management, Computer Science and Communications Research Centre, Polytechnic Institute of Leiria, P-2411-901, Leiria, Portugal
e-mail: filipe.felisberto@ipleiria.pt

Isabel Marcelino · António Pereira
INOV INESC INOVAÇÃO – Instituto de Novas Tecnologias Leiria, Portugal
e-mail: apereira@ipleiria.pt

Florentino Fdez-Riverola
ESEI: Escuela Superior de Ingeniería Informática, University of Vigo, Edificio Politécnico, Campus Universitario As Lagoas s/n, 32004, Ourense, Spain
e-mail: riverola@uvigo.es

P. Novais et al. (Eds.): Ambient Intelligence - Software and Applications, AISC 92, pp. 85–92.
springerlink.com

that the percentage of people older than 65 years old will raise from 17,1% to 30%. This corresponds to an increase of 84.6 million people in 2008 to 151.1 million people in 2060[7].

With the aging problems in mind, the Elder Care project was envisioned [10]. Elder Care distinguishes itself from other health care projects, by providing a complete package for elder health care support. It gives complete monitoring support of the elderly not only in his home, but also when he is away from it. This project does not limit its concerns to the physical health of the elderly, it also has modules responsible for making sure that the mental health of the elderly does not deteriorate due to loneliness and social neglecting.

To be able to provide this type of support, Elder Care is divided in various sub-projects, being one of them the Body Monitor module, that has the intention of developing a system to allow the constant monitorization of the elderly's vital signs. This module is only possible due to the big advancements in the domain of miniaturization technology, the emergence of wireless sensor technologies and more recently the studies made in the area of Body Area Networks[8]. One of the items in study by the Body Monitor module is fall detection, and it is where our atention is mainly focused in this article.

In a recent report from the Health Evidence Network for the World Health Organization [12], it is established that 30% of the population over the age 65 falls at least once each year, this percentage is even higher in the group of 80 and over, where it reaches 50%. Still not only falls are the main reason for injuries that need hospitalization, but it is also the primary cause of injury-related deaths.

Not only there is a problem with the actual impact, but also if the fall is unattended for a long period of time, will by itself, bring additional problems to the elderly. The contact with the ground may lower the body's temperature to values that can lead to delicate health conditions, like for example pneumonia [6]. The longer the elderly stays in the ground unattended, more will he be prone to become fearful of having a normal day-to-day life [12].

In this article, the Body Monitor architecture is being approached and the fall detection tests exposed. Our main concern is to show the evolution of our tests that aim to reduce, and ultimately eliminate,false positives. Taking into account the constraints underlying sensor's energy costs.

2 Fall Detection

Many studies have been done on the area of computer assisted fall detection; in the early 90' with the advent of more accurate sensors, and after their early project using video monitoring failed due to legal and moral concerns, the researchers Lord and Colvin using a small tri-axial analog accelerometer started studying the acceleration sustained by the human body during the impact [9]. This approach was the one followed by the majority of the following studies. Some years later, in 1998, was developed the first prototype of a fall detection device to be used on a telecare system. Like the Lord and Colvin study, this device detected the fall by measuring the force of the impact when the body hit the ground, and complemented it by using a mercury sensor to verify if the person was on a laying position [13].

This first systems all had a sub 90% success rate, and suffered from many false positives, this was due in part to the inefficient hardware being used and to the way the problem was being approached, as the system relied on the energy of the impact to be transmitted through the human body, in case the energy was absorbed by the body the fall could not be detected or would be confused with an activity of daily living (ADL).

In order to mitigate this problem, a study was conducted by Ge Wu [14]. In this study, instead of focusing on the actual impact, Wu decided to study the velocity of the human body during both ADL and falls, using image processing equipment he measured the velocity of the body during multiple types of activities that would normally cause false positives. After studying the results, he discovered clear differences between ADL and falls, for example, both the vertical and horizontal velocities are between two and three times higher during a fall than during a normal activity. Not only did this study prove, that it was more precise to use velocity instead of only the impact, it also proven that by using velocity it was possible to identify a fall even before the impact occurs.

More recent works [3, 5, 4], by Alan K. Bourke, have taken the study of ADL differentiation from falls a step further, in [3] it is proven that it is not necessary to use both vertical and horizontal velocities to undoubtedly detect a fall. This works have also proven that it is not necessary to measure the velocity of multiple parts of the human body, because the velocity of the trunk by itself, is enough to correctly identify a fall.

3 Body Monitor Architecture

The research team behind Body Monitor has decided to implement a scalable architecture for Body Area Networks, this enables that in the future other types of sensors, not only those chosen for the fall detection, can be easily added to the system. The proposed architecture consists in a set of sensor nodes positioned on a human body, collecting data and detecting abnormalities. The information is afterwards sent to a central sensor node (nominated by coordinator). This coordinator will then send information to a base station (Root) or a mobile phone (PDA). The base station or mobile phone will manage alerts sent according to the occurred events, see fig. 1.

One of the biggest problems in a Wireless Network, like this one, is the energy spent on the communication. To optimize the power consumption this architecture uses a routing protocol that tries to forward data with the minimum hops possible.

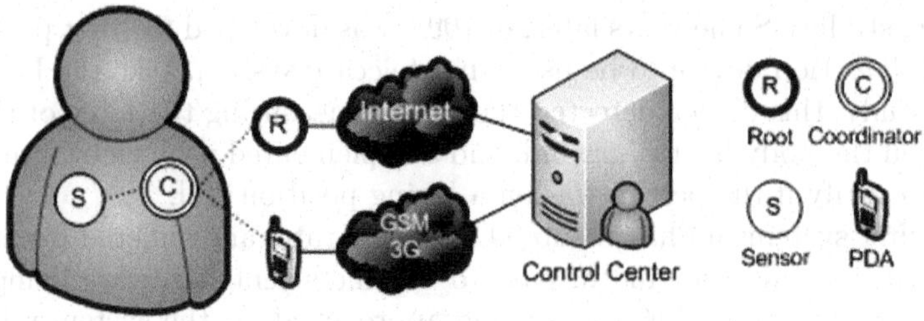

Fig. 1 The Body Monitor Architecture

In terms of the actual fall detection system, it is not just a simple process of making decisions based on individual data, obtained from the accelerometer sensor. This type of solution would create a lot of false positives, which by itself is nefarious for the acceptance of the project, because it would create a "peter and the wolf" problem where the warnings would stop to be taken in account and when a real fall occurred it would be ignored. Another problem would be the battery drain caused by all the connections made to the central system.

So to implement a viable solution, it is necessary to be able to distinguish a real fall from an ADL, with such a system in place it has been proven that a fall can be detected and confirmed even before the actual impact occurs [14]. It is also important to take in account that this is a BAN so the system should have the least energy consumption possible, to do this it is necessary to study what is the lowest sampling rate necessary to guarantee a near 100% success rate on fall detection without causing false positives.

For this system a tri-axial accelerometer has been selected, each one of its axis is positioned perpendicular to the other, simulating the orthogonal axis. The data collected from this sensors is then stored, so that it can be continuously analyzed. Also, the values of each axis are analyzed independently and in conjunction, with this it is hoped not only to be able to detect and distinguish unhampered falls, but also to reinforce the data gathered from other sensors. For example, a decrease of the elderly's activity followed by an immobilization, if accompanied by an abnormal variation of the heart rate, would undoubtedly be a cause for an alert.

4 Tests

The first testing stage consisted on implementing the detection model on a prototype board. This research group has opted to use the iMote2 platform [11] instead of an Atmel-based platform. The iMote2, with its PXA271 XScale Processor, is able to achieve processing speeds of up to 520MHz [1], a lot higher than the commonly used Atmel ATmega128L microcontroller can with its 8MHz processor[2]. At these clock rates, it was expected that the power consumption would also be many times superior, but this is avoided due to the scalability of the PXA271 processors, that allows them to work at speeds as low as 13MHz; at these speeds its possible to obtain power consumptions near the ones of the Atmel processor, while still being able to manage more sensor sources. In particular, the iMote2 is able to acquire, process and send data from three acceleration axis without stopping the sampling cycle. For testing purposes, the accelerometer used was the one present on the Crossbow iMote2 sensor board, the ST Micro LIS3L02DQ.

To be able to correctly test the prototype, it was first necessary to correctly calibrate the sensors and prepare the system for the acceleration data being returned. This is very important because the accelerometer sensors do not only return acceleration when there actually occurs movement, they are always returning the acceleration applied to them by the gravity field. It was also important, to gather data of a fall when there was only the force of gravity in action.

To accomplish this, was necessary to design a second testing phase with as few as possible external interferences, so it was decided not to use any human subjects in this phase. In order to study a vertical fall it was used a 5kg box with the sensor coupled to it, this box was then dropped from a height of one meter. To test a fall where there were more axis involved that just the vertical axis, was design a system that would create falls with a consistent arc trajectory. This system consists on two wood boards, one of them with the size of 1,7m to simulate the height of a human being, the size of the other board is irrelevant as it only serves to fixate the system to the ground; the one representing the human is placed upright and are both connected on the extremities by a hinge. During the actual test the sensor was placed in one of two positions, at a high of 0,85m to simulate the height of the hip and at 1,45m to simulate the height of the shoulder.

Following the calibration of sensors and the creation of a reference table with the values from the previous tests, comes the third testing phase that consists on the study of the ADL of a voluntary. As it is not practical to record the complete day of a person and then evaluate the results, due to the immense data that would be obtained, some simple and direct tasks are performed instead. The tasks performed by the voluntary are:

- Walk in both directions on an L-shaped route
- Sit and raise from a chair
- Pick an object from the ground, in the correct way (by bending the knees) and the wrong way (by bending the back)
- Go up and down a flight of stairs, both in a slow pace and in a quick one

The forth and last phase, consist on testing actual falls using young voluntaries in a controlled environment. This final test, was used to validate the gathered information and to confirm that if with even more noise it was still possible to correctly detect a fall.

5 Analysis of Results

During the ADL tests what was analyzed, was that none but the "sit and raise from a chair" test, presented any data that could be confused with the one collected from the previous fall tests. In fourth phase, with tests of normal falls, the acceleration was enough to detect the fall and there was not any problem in distinguishing it from an ADL, a representation of the data collected can be seen in fig. 2.

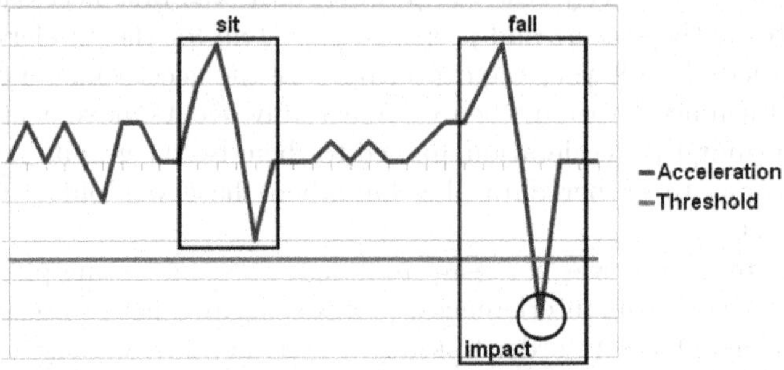

Fig. 2 Representation of a test using only acceleration

In the case of the sit down tests, when the volunteer would abruptly let is body fall on the chair, the data collected had samples of the same magnitude of those of an actual fall. This occurs because, during a more brusque sit down, the upper part of the body is not sported so its, almost, in free fall.

The problems also appears when the fall is not completely unhampered, and the person is able to hold to something before actually hitting the ground, this causes an intermediary deceleration causing the final impact to be of a smaller magnitude. This in conjunction with the human body's natural absorption of the impact, can lead to problems in differentiating the fall acceleration values from those of a sit down action done with more violence, see figure 3.

When using the velocity instead of the acceleration, the small intermediary deceleration would not have that much impact on the final results, as the speed of the person is the result of the combination of past and present data. The data gathered in velocity is a stream of continuous values, and not a group of isolated values like in the case of acceleration. The fall test in figure 4 is a representation of the same data as in figure 3, but this time using velocity.

Unfortunately, using velocity instead of acceleration has its own perils, while the use of acceleration is a direct analyze of the data gathered from the sensors, the velocity must be calculated, so it needs extra processing power and it is more sensitive to loss or errors in the gathered data.

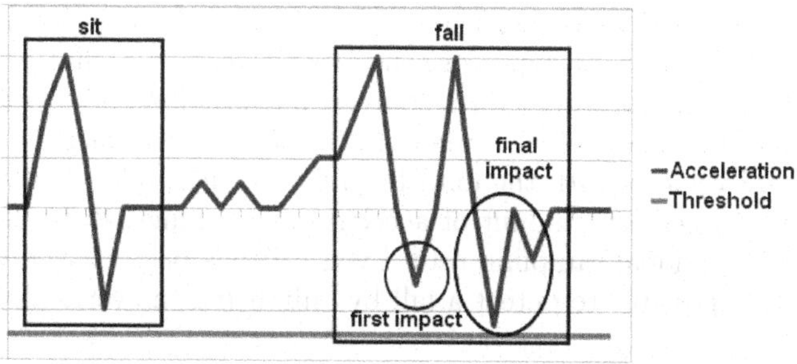

Fig. 3 Problem of only using acceleration

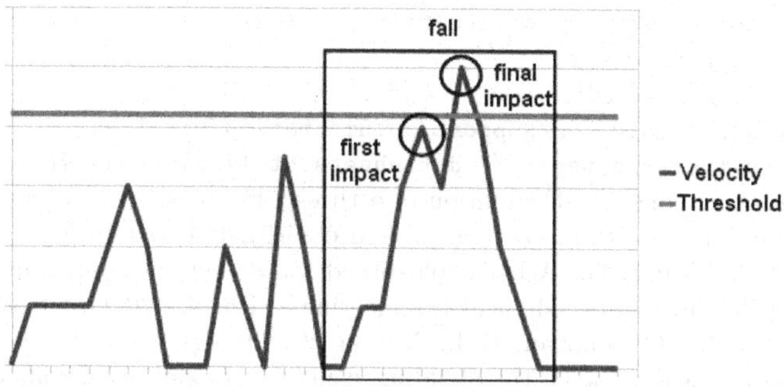

Fig. 4 Detection using velocity

6 Conclusions and Future Work

Detecting a fall using wireless sensor networks has two major problems: distinguishing false positives and sparing sensor battery.

It is very important to identify unequivocally situations where a fall happens and situations where other human movements are made and can be wrongly interpreted as falls. Otherwise, alerts will never be reliable. It is also very important, when working with wireless sensors, to have in mind their limitations, especially battery limitations. A continuous communication may lead to the crash of the system in a few hours.

For the main function of the system, the detection of unhampered falls, instead of relying on the older technique of using the body's impact, it has been decided to use the calculated velocity. This decision came after studying, the already referenced works of Wu and Bourke, and after confirming it with the available equipment. But using velocity over acceleration may overload the sensors and crash the system. So it is important to gather not only more information about acceleration and / or velocity, but also parameters such as abnormal variation of the heart rate or even interpret frames from a camera monitoring the elderly.

Regarding the fall detection using only physical data sensors, there are already plans to test new types of sensors and verify to what extent, it is beneficial the addition of new data sources, like oscilloscopes, to the energy consumption. It should also be evaluated, that if by combining the data from the horizontal velocity with the data from the vertical velocity it is possible to reduce the necessary amount of acceleration readings, thus allowing the reduction of the global sampling rate. Even if it has been proven by Bourke in [5] that it is possible to detect a fall by only using the vertical velocity.

References

1. Imote2 datasheet. PDF, http://ubi.cs.washington.edu/files/imote2/docs/imote2-ds-rev2.0.pdf
2. Mica2 datasheet. PDF, http://www.memsic.com/support/documentation/wireless-sensor-networks/category/7-datasheets.html?download=147
3. Bourke, A., Lyons, G.: Evaluation of a threshold-based tri-axial accelerometer fall detection algorithm. Gait and Posture 26(2), 194–199 (2007)
4. Bourke, A., Lyons, G.: A threshold-based fall-detection algorithm using a bi-axial gyroscope sensor. Medical Engineering & Physics 30(1), 84–90 (2008)
5. Bourke, A.K., O'Donovan, K.J., Nelson, J., OLaighin, G.M.: Fall-detection through vertical velocity thresholding using a tri-axial accelerometer characterized using an optical motion-capture system, pp. 2832–2835 (2008)
6. Thorn, G., Adams, R., Braunwald, E., Isselbacher, K., Petersdorf, R.: Harrison: Principios de medicina interna, 5a edn. McGraw Hill Interamericana, Mexico (1981)
7. Giannakouris, K.: Ageing characterises the demographic perspectives of the European societies. Eurostat: Statistics in Focus. Retrieved 9, 8–72 (2009)
8. Karulf, E.: Body area networks (ban). pdf (2008), http://www.cse.wustl.edu/~jain/cse574-08/ftp/ban.pdf
9. Lord, C., Colvin, D.: Falls in the elderly: Detection and assessment, pp. 1938–1939 (1991)
10. Marcelino, I., Barroso, J., Cruz, J.B., Pereira, A.: Elder care architecture. In: Proceedings of the 2008 Third International Conference on Systems and Networks Communications, pp. 349–354 (2008)
11. Nachman, L., Huang, J., Shahabdeen, J., Adler, R., Kling, R.: Imote2: Serious Computation at the Edge, pp. 1118–1123 (2008)
12. Todd, C., Skelton, D.: What are the main risk factors for falls among older people and what are the most effective interventions to prevent these falls? Tech. rep., WHO Regional Office for Europe (Health Evidence Network report) (2004)
13. Williams, G., Doughty, K., Cameron, K., Bradley, D.: A smart fall and activity monitor for telecare applications 3, 1151–1154 (1998)
14. Wu, G.: Distinguishing fall activities from normal activities by velocity characteristics. Journal of Biomechanics 33, 1497–1500 (2000)

Multi-Agent System for Detecting Elderly People Falls through Mobile Devices

Patricia Martín, Miguel Sánchez, Laura Álvarez,
Vidal Alonso, and Javier Bajo

Abstract. Falls in the elderly and disabled people represent a major health problem in terms of primary care costs facing the public and private systems. This paper presents a multi-agent system capable of detecting falls through sensors in a mobile device and act accordingly at runtime. The new system incorporates a fall detection algorithm based on machine learning and data classification using decision trees. The base of the system are three types of interrelated agents that coordinate to know the position of a user from data obtained through a mobile terminal, and GPS position, which in case of fall may be sent via SMS or by an automatic call. The proposed system is self-adaptive, since as new fall date is incorporated, the decision mechanisms are automatically updated and personalized taking into account the user profile.

Keywords: Context-Aware, Multi-Agent Systems, Decision trees.

1 Introduction

Falls are one of the most serious problems in the geriatric care because they are one of the fundamental causes of injury and even death in elderly people. According to the World Health Organization (WHO), between 28% and 34% of people 65 and older experience at least one fall per year [1]. In Spain the data are also alarming. The on mortality of the Institute of Health Carlos III in 2007 reveal that the greater the age the higher the mortality rate for accidental falls, exceeding even the percentage of some types of cancer. Moreover, various authors [1, 2, 3, 4, 5, 6, 7], consider that falls are due to a combination of multiple factors, and can be

Patricia Martín · Miguel Sánchez · Laura Álvarez · Vidal Alonso · Javier Bajo
Universidad Pontificia de Salamanca, c/ Compañía 5,
37002 Salamanca, Spain
e-mail: `patrimrodilla@usal.es`, `miguel@chocosoft.net`,
`lalvarezba@gmail.com`, `{valonsose,jbajope}@upsa.es`

P. Novais et al. (Eds.): Ambient Intelligence - Software and Applications, AISC 92, pp. 93–99.
springerlink.com

grouped into two main groups: intrinsic and extrinsic factors, and consequences of they are not only physical but also psychological, social and economic [2, 8, 9].

This situation makes necessary to investigate in new solutions aimed at fall's detection in an effective an non-intrusive manner. This paper proposes an innovative multi-agent system specifically designed to detect falls and act accordingly. The core of the system is a detection mechanism built into a mobile phone. This mechanism allows a simple and non-intrusive detection of falls, and enables ubiquitous communication mechanisms.

The reminder of the paper is structured as follows: First we review the state of the art of the technologies involved in the work. In section 3 the proposed multi-agent system is presented. Finally, Section 5 presents the results and conclusions obtained after applying the architecture to a real case study.

2 Related Work

After consulting the literature, there not exists a multi-agent system aimed at the detection of falls of elderly people making use of mobile devices for detecting the fall. Different researchers have tackled the problem from different perspectives, and there are three elements that give an innovative character to this project: i) the focus on the mobile device, which provides a certain level of independence to the physical detection component, ii) the multi agent system, that provides ubiquitous distributed solving abilities and learning capacity, and iii) the fact that the system is able to perform additional functionalities within the field of e-Health and care of the elderly.

A study of the related work reveals that there exist projects aimed at designing multi-agent system for the detection of falls at home [11], with agents located in a central computer connected to a health center and various sensors placed around the house, and a device specially built for this purpose. The above solution is focused in the field of home automation, and as other specific approaches [18] [20] does not provide additional functionality (the mobile phone itself) to the patient. On the other hand, there are specific projects aimed at finding an effective algorithm for the detection of falls using three-axis accelerometer [13] [16], gyroscopes [17] [13], or a combination of both [12]. These systems have evolved during the last years, because they were initially conceived to detect falls or collisions. The current systems try to detect the problems in advance (pre-impact detection) [15] [12]. Some of the existing approaches use the threshold as the key component of the detection algorithm. The threshold is the sharp difference of gravity between two axis of the accelerometer. The approach proposed in this paper makes use of the threshold, but it is not a fundamental component. It is necessary to note that these solutions do not use agents for decision making, and there is not standardized solution. Some authors focus on semi-supervised learning techniques to increase the accuracy of fall detection, making use of video images for posture recognition [21], and even sounds [19] [22] or extraction of 3D trajectories of the body or body parts [23].

The system presented in this paper makes use of intelligent agent to improve its learning abilities. The agents make use of the measurements of two routine medical test (get up and go, get & go), as well as of the patient data. The main novelties of the approach presented in this paper are the distributed problem solving abilities of the multi-agent system, the use of a mobile phone as a fall detector, that can accurately detect whether there is a fall and alert and the medical center or caregiver more close, without limiting the rest of the functionalities of the phone. In addition, the system covers other needs of elderly users, as reminders of appointments or taking medication.

3 Proposed Architecture

The proposed approach primarily based on harnessing the benefits of multi-agent systems in the problem of fall detection. A multi-agent system can notably help to make automatic decisions and to integrate advanced detection mechanisms into mobile phone devices. Intelligent agents are very appropriated to be installed into mobile devices, and given the increasing relevance that this kind of devices are acquiring in our society, sensing and making decisions from mobile devices can notably help to develop ubiquitous, un-obstructive and intelligent environments. With the current inter-connection possibilities, the decision may be taken from the device itself or remotely, based on a more complex analysis of the information. The approach makes use of the phone's functionalities to manage an automatic alert system, connected to a care center that takes advantage of the GPS system to provide accurate information about the patient location. The architecture of the proposed multi-agent system is composed of three agent types that collaborate to detect falls: Sensor agent, Classifier agent and Actuator agent. The structure of the multi-agent architecture can be observed in Figure 1. The agent types are explained in detail in the following sub-sections.

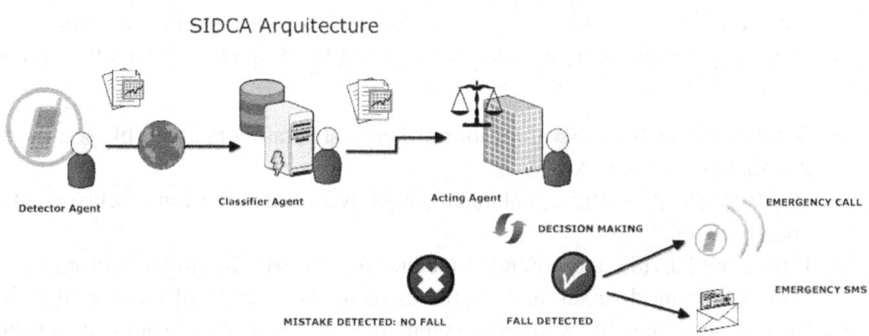

Fig. 1 Multi-agent architecture to manage fall detection

3.1 Sensor Agent

It is an agent specially designed to be installed on mobile devices. It acts as an interface between the user and the rest of the system and has advanced capabilities for fall detection. The Sensor agent uses the accelerometers of a mobile phone to collect data of the movements of the elderly person carrying the phone. In this way, the agent gets the raw data that is used to obtain a classification and identify falls. This lightweight classification process is based on the use of previous experiences: the system evaluates similarity to previously stored patterns. If the result of the classification is not accurate, then the agent communicates with the Classifier agent to carry out an advanced classification.

It is necessary to take into account that the mobile device must be equipped with three-axis accelerometer (X, Y, Z) on a single silicon chip, including in that the electronics that processes the signals. These kinds of accelerometers are currently available in most of the terminals on the market. Thanks to this technology, the Sensor agent located in the mobile can manage the three main values required for the classification process: rotation, translation and space and temporal situation of the mobile device. An initial position is taken as a reference. The agent identifies eight different patterns: standing, walking, climbing or descending stairs, sitting or to detect fall forward and fall back. It is also necessary to take in mind that the system uses the JADE Leap suite for Android. This technology allows us to integrate the Sensor agent into the multi-agent system.

3.2 Classifier Agent

The classifier agent is located in the care center and manages the data recorded by the Sensor agent. The Classifier agent implements a decision algorithm based on the J48 decision tree. The agent learns progressively and dynamically recalculates the values that are given to the position of the falls for that particular user. This agent is necessary, since it is in charge of personalizing the fall patterns to the user profile. The profile takes into account the motion of each elder or dependent person, and the concrete peculiarities. Basically the decision-making mechanism is based on J48 decision trees, an adaptation of the C4.5 algorithm, which is common for classification. Basically, the agent:

- Works with continuous values for attributes, separating the possible results in 2 branches Ai<=N y Ai>N.
- The trees are less dense, and each sheet covers a distribution of classes and not a particular class.
- It uses the "divide and conquer" method to generate the initial decision tree from a training data set, and continues recursively. Each of the iterations is carried out when the Classifier agent receives new data from the Sensor agent.
- Implements a strategy based on the use of the gain ratio criteria, defined as I (Xi, C) / H (Xi). Thus it is possible to avoid potential benefits to the selection of those variables with the greatest number of possible values.

Once the Sensor agent has classified the new input data, it sends the border values calculated for the decision tree, together with the input data, to the Actuator agent.

3.3 *Actuator Agent*

The actuator agent is located in the care center and stores all the information captured by the Sensor agent and processed by the Classifier agent. With this information, the agent knows if the user has suffered a fall or not, and whether there has been an anomaly in the pattern of falling for the user. The actuator agent decides the action required for the emergency. The agent is also in charge of registering false positives and informs the Sensor and Classifier agents about these situations. When the Actuator agent receives information about a fall detection, it executes two actions:

- The mobile terminal performs an automatic emergency call to the emergency number. It also detects the closest medical center or the emergency number more appropriate depending on the user profile.
- The mobile terminal automatically sends an SMS to the contact person configured by the user. This SMS contains the GPS position of the mobile terminal, and information about the potential fall details.

4 Results and Conclusions

The system was tested under simulation conditions. The test's parameters were set up using previous postural studies related to elderly people. Over three months of testing 500 files were obtained using data from real falls. The Sensor agent received the input data and processed it into XML (eXtensible Markup Language), keeping this format for the rest of the agents. An example of XML file is shown in Figure 2. As can be seen in Figure 2, stores the data obtained by the accelerometer: time, x, y and z values, pitch and roll.


```xml
<?xml version="1.0" encoding="UTF-8" ?>
<!DOCTYPE Sidca SYSTEM "SIDCA.dtd">
<vector time="2010-05-06 10:31:52.515">
        <xyz>-2.9147544 , 5.053149 , 7.3141265 , </xyz>
        <pitch>78.43</pitch>
        <roll>34.95</roll>
</vector>
<vector time="2010-05-06 10:31:52.609">
        <xyz>-2.6014864 , 4.944186 , 7.273266 , </xyz>
        <pitch>45.29</pitch>
        <roll>12.43</roll>
</vector>
<vector time="2010-05-06 10:31:52.609">
        <xyz>-2.4516625 , 4.862464 , 7.3958488 , </xyz>
        <pitch>-148.33</pitch>
        <roll>-0.87</roll>
</vector>
<vector time="2010-05-06 10:31:52.609">
        <xyz>-2.152015 , 5.284695 , 7.3141265 , </xyz>
        <pitch>178.90</pitch>
        <roll>-23.83</roll>
</vector>
```


Fig. 2 Example of XML file used by the agents in the system

During three months the system was trained and tuned trying to obtain a good learning rate. The system can successfully distinguish different positions (and five types of falls): standing, walking, climbing or descending stairs, sitting or to detect fall forward and fall back. The system is able to recalculate the parameters of the user in real-time. The error rate decreases as the system learns from the new input data, and the minimum mean quadratic error obtained in the experiments was 0,16.

The main advantages of the proposed approach can be summarized as follows:

- The computational cost is minimal. The algorithm for the mobile terminal is very light and causes a low battery consumption.
- The reliability of data processing in the three agents is acceptable.
- The traffic generated into the network is acceptable and very appropriated for distributed systems of this kind.
- The system provides a mechanism to reduce the impact of the falls for elderly people, providing and automatic and quick response to the potential incidents.
- The system provides financial benefits. The economic consequences of falls in elderly people is very costly to health services [2] because they must meet various demands such as: the initial hospitalization after a fall with serious consequences, treatments and surgical and orthopedic the need for a further period of rehabilitation to try to restore the patient to their prior functional status, and extra care of elderly hospital with minor injuries and both caregivers and institutionalization costs.

The results obtained in the experiments are promising, but need to be improved using a more extensive dataset and evaluating the approach in real scenarios. Our future work focuses on obtaining more specialized algorithms for classification and learning.

Acknowledgements. This study has been supported by the Spanish Ministry of Science and Innovation project TRA2009_0096.

References

1. Suelves, J.M., Martínez, V., Medina, A.: Lesiones por caídas y factores asociados en personas mayores de Cataluña, España. Rev. Panam Salud. Publica. 27(1), 37–42 (2010)
2. Carro García, T., Alfaro Hacha, A.: Caídas en el anciano. Residentes de Geriatria Hospital Virgen de Valle. Toledo
3. Quiénes son ancianos frágiles – ancianos de riesgo? Estudio en personas mayores de 65 años del Área Sanitaria de Guadalajara
4. Esmeralda, M.R., et al.: Incidencia de caídas en la Unidad de Hemodiálisis del Hospital General de Vic (Barcelona). Rev. Soc. Esp. Enferm. Nefrol. 11(1), 6469 (2008)
5. Prat, I., Fernandez, E., Martinez, S.: Detección del riesgo de caídas en ancianos en atención primaria mediante un protocolo de cribado. In: 2007 Área Básica de Salud de Palamós. Serveis de Salut Integrats Baix Empordà. Palamós. Girona. España (2007)

6. Papiol, M., Duaso, E., RodríguezCarballeira, M., Tomás, S.: Identificación desde un servicio de urgencias de la población anciana con riesgo de caída que motiva ingreso hospitalario. In: Servicio de Urgencias y Unidad Funcional de Geriatría. Hospital Mutua de Terrassa, Barcelona (2003)
7. Sociedad española de enfermería de urgencias y Emergencias. Prevención de caídas. Recomendación científica 10/05/10 de 25 de junio de (2009)
8. da Gama Silva, Z.A., Gomez, A., Sobral, M.: Epidemiología de caídas de ancianos en España: Una revisión sistemática. Rev. Esp. Salud Publica. 82(1), 4355 (2007)
9. Lázaro, A.: Características de las caídas de causa neurológica en ancianos
10. Lázarodel Nogal, M., LatorreGonzález, G., GonzálezRamírez, A., RiberaCasado, J.M.:
11. Pan, J.I., Yung, C.J., Liang, C.C., Lai, L.F.: An Intelligent Homecare Emergency Service System for Elder Falling. In: Proceedings of IFMBE World Congress on Medical Physics and Biomedical Engineering, vol. 14 (2006)
12. Nyan, M.N., Tay, F.E.H., Murugasu, E.: A wearable system for preimpact fall detection. Journal of Biomechanics 41, 3475–3481 (2008)
13. Bourke, A.K., Lyons, G.M.: A thresholdbased falldetection algorithm using a biaxial gyroscope sensor. Medical Engineering & Physics 30(1), 8490 (2008)
14. Bourke, A.K., O'Brien, J.V., Lyons, G.M.: Evaluation of a thresholdbased triaxial accelerometer fall detection algorithm. Gait & Posture 26(2), 194–199 (2007)
15. Bourke, A.K., O'Donovan, K.J., ÓLaighin, G.: The identification of vertical velocity profiles using an inertial sensor to investigate preimpact detection of falls. Medical Engineering & Physics 30(7), 937–946 (2008)
16. Kumar, A., Rahman, F., Lee, T.: IFMBE. In: Proceedings: 13th International Conference on Biomedical Engineering ICBME 2008, Singapore, December 3–6 (2009)
17. Lindemann, U., Hock, A., Stuber, M., Keck, W., Becker, C.: Evaluation of a fall detector based on accelerometers: A pilot study. Medical and Biological Engineering and Computing 43(5) (October 2005)
18. Lustrek, M., Kaluza, B. (2009) Fall detection and activity recognition with machine learning. Slovenian Society Informatika, report of (May 2009)
19. Zhang, T., Wang, J., Xu, L., Liu, P.: Fall Detection by Wearable Sensor and OneClass SVM Algorithm. Intelligent Computing in Signal Processing and Pattern Recognition 345 (2006)
20. Doukas, C., Maglogiannis, I.: Advanced patient or elder fall detection based on movement and sound data. Pervasive Computing Technologies for Healthcare (2008)
21. Sixsmith, A., Johnson, N.: A Smart Sensor to Detect the Falls of the Elderly, vol. 3(2), pp. 42–47. IEEE Computer Society, Los Alamitos (2004)
22. Londei, S.T., Rousseau, J., et al.: An intelligent videomonitoring system for fall detection at home: perceptions of elderly people. Journal of Telemedicine and Telecare 15(8), 383–390 (2009)
23. Zigel, Y., Litvak, D., Gannot, I.: A Method for Automatic Fall Detection of Elderly People Using Floor Vibrations and Sound—Proof of Concept on Human Mimicking Doll Falls. IEEE Transactions on Biomedical Engineering 56(12), 2858–2867 (2009)
24. Rougier, C., Meunier, J.: Demo: Fall Detection Using 3D Head Trajectory Extracted From a Single Camera Video Sequence. Journal of Telemedicine and Telecare 11(4) (2005)

Intelligent Video Monitoring for Anomalous Event Detection

Iván Gómez Conde, David Olivieri Cecchi, Xosé Antón Vila Sobrino,
and Ángel Orosa Rodríguez

Abstract. Behavior determination and multiple object tracking for video surveillance are two of the most active fields of computer vision. The reason for this activity is largely due to the fact that there are many application areas. This paper describes work in developing software algorithms for the tele-assistance for the elderly, which could be used as early warning monitor for anomalous events. We treat algorithms for both the multiple object tracking problem as well simple behavior detectors based on human body positions. There are several original contributions proposed by this paper. First, a method for comparing foreground - background segmention is proposed. Second a feature vector based tracking algorithm is developed for discriminating multiple objects. Finally, a simple real-time histogram based algorithm is described for discriminating movements and body positions.

Keywords: computer vision, foreground segmentation, object detection and tracking, behavior detection, tele-assistance, telecare.

1 Introduction

Life expectancy worldwide has risen sharply in recent years. In 2050 the number of people aged 65 and over will exceed the number of youth under 15 years, according to recent demographic studies [7]. Combined with sociologic factors, there is thus a growing number of elderly people that live alone or with their partners. While people may need constant care, there are two problems: not enough people to care for elderly population and the government can not cope with this enormous social spending. Thus, Computer Vision can provide a strong economic savings by eliminating the need for 24 hour in-house assistance.

Computer vision has entered an exciting phase of development and use in recent years. Present applications go far beyond the simple security camera of a decade

Iván Gómez Conde · David Olivieri Cecchi ·
Xosé Antón Vila Sobrino · Ángel Orosa Rodríguez
University of Vigo (Department of Computer Science)
e-mail: `{ivangconde,olivieri,anton,aorosa}@uvigo.es`

P. Novais et al. (Eds.): Ambient Intelligence - Software and Applications, AISC 92, pp. 101–108.
springerlink.com

ago and now include such fields as assembly line monitoring, robotics, and medical tele-assistance. Indeed, developing a system that accomplishes these complex tasks requires coordinated techniques of image analysis, statistical classification, segmentation and inference algorithms.

The motivation for this paper is the development of a tele-assistance application, which represents a useful and very relevant problem domain. First we must detect what we consider to be the foreground objects [4, 5], we must then track these objects in time (over serveral video frames) [8] and discerning something about what these objects are doing. There is a large body of literature in the area of human action recognition. For segmentation and tracking, for example, the review by Hu [3] provides a useful review and taxonomy of algorithms used in multi-object detection. For human body behavior determination from a video sequences, the recent reviews by Poppe [6] and Forsyth [2] provide a whirlwind tour of algorithms and techniques.

In this paper, we describe the architecture of our software system, as well as details of motion detection, segmentation of objects, and the methods we have developed for detecting anomalous events. Finally, we show the performance results and conclusions of this work.

2 The Software System

Our software application has been written in C++ and uses the OpenCV library [1], which is an open-source and cross-platform library, for developing a wide range of real-time computer vision applications. OpenCV implements low level image processing as well as high level machine learning algorithms. For the graphical interface, the QT library is used since it provides excellent cross-platform performance.

This software is an experimental application, the graphical interface is designed to provide maximum information about feature vector parameters. Thus, the system is not meant for end-users at the moment. Instead, the architecture of the system provides a plugin-framework for including new ideas. A high level schema of our software system is shown in Figure 1 with the component diagram.

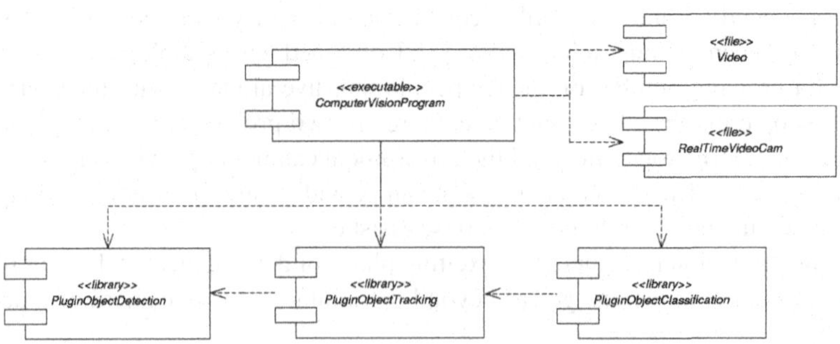

Fig. 1 Computer Vision system for video surveillance

2.1 Foreground Segmentation

The first phase of extracting information from videos consists of performing basic image processing: loading the video, capturing individual frames, and applying various smoothing filters. Next, blobs are identified based upon movement between frames. There are several background subtraction methods: Running Average and Gaussian Mixture Model (Figure 2).

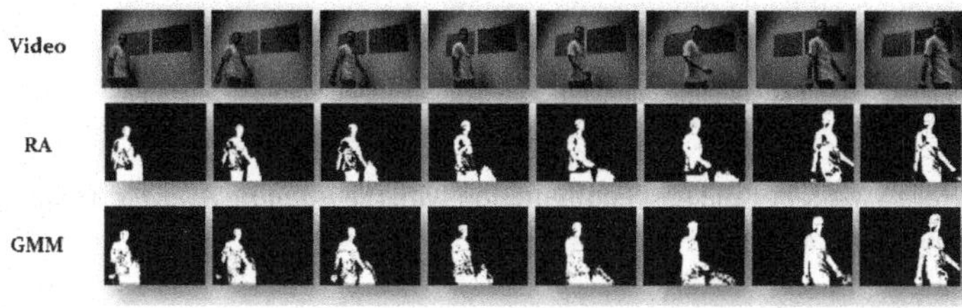

Fig. 2 Execution of "Running Average (RA)" and "Gaussian mixture model (GMM)"

The **Running Average** [1] is by far easiest to comprehend. Each point of the background is calculated as by taking the mean of accumulated points over some pre-specified time interval, Δt. In order to control the influence of previous frames, a weighting parameter α is used as a multiplying constant in the following way:

$$A_t(x,y) = (1-\alpha)A_{t-1}(x,y) + \alpha I_t(x,y) \tag{1}$$

where the matrix A as the accumulated pixel matrix, $I(x,y)$ the image, and α is the weighting parameter. We have tested 8 executions with values of α betweeen 0 and 0.8.

The **Gaussian Mixture Model** [4] is able to eliminate many of the artefacts that the running average method is unable to treat. This method models each background pixel as a mixture of K Gaussian distributions (where K is typically a small number from 3 to 5). The probability that a certain pixel has a value of x_N at time N can be written as:

$$p(x_N) = \sum_{j=1}^{K} w_j \eta(x_N; \mu_j, \sigma_j^2) \tag{2}$$

where w_j is the weight parameter of the j_{th} Gaussian component, and $\eta(x_N; \mu_j, \sigma_j^2)$ is the Normal distribution of j_{th} component.

The K distributions are ordered based on the *fitness value* $\frac{w_j}{\sigma_j}$ and the first B distributions are used as a model of the background of the scene where B is estimated as:

$$B = \arg\min \left(\sum_{j=1}^{b} w_j > T \right) \tag{3}$$

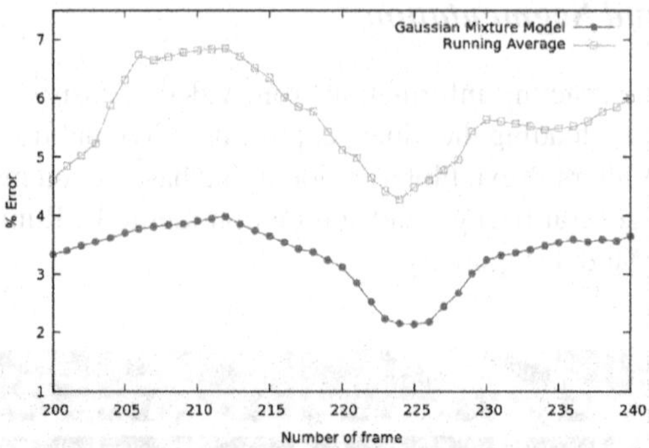

Fig. 3 % error ($\frac{FN+FP}{640 \cdot 480}$) of the best configuration with the *Running Average Model* and with the *Gaussian Mixture Model*

The threshold T is the minimum fraction of the background model. In other words, it is the minimum prior probability that the background is in the scene. In this paper we have tested 8 executions with different values of variance between 1 and 5.

The algorithms *Running Average* and *Gaussian Mixture Model* have been tested with our computer vision system. The data consists of 1 video sequence of resolution 640x480 pixels, 22 seconds of duration and 25 frames per second. We select the frames between 200 and 240. For each frame, it was necessary to manually segment foreground objects in order to have a *ground truth* quantitative comparison. We calculate the number of foreground pixels labeled as background (false negatives - FN), and the number of background pixels labeled as foreground (false positives - FP), and the total percentage of wrongly labeled pixels $\frac{FN+FP}{640 \cdot 480}$. The figure 3 shows the best results for the *Running average* with small values for alpha $(\alpha = 0.05)$ and for the *Gaussian mixture model* with $(\sigma = 2.5)$.

2.2 Finding and Tracking Individual Blobs

Foreground objects are identified in each frame as rectangular blobs, which internally are separate images that can be manipulated and analyzed. In order to classify each blob uniquely, we define the following feature vector parameters: (a) the size of the blob, (b) the Gaussian fitted values of RGB components, (c) the coordinates of the blob center, and (d) the motion vector. The size of blobs is simply the total number of pixels. Histograms are obtained by considering bin sizes of 10 pixels. We also normalize the feature vectors by the number of pixels.

In order to match blobs from frame to frame, we perform a clustering. Since this can be expensive to calculate for each frame, we only recalculate the full clustering algorithm when blobs intersect. Figure 4 shows excellent discrimination by using the norm histogram differences between blobs for each color space. The x-axis is the norm difference of red, while the y-axis is the norm difference histogram for green.

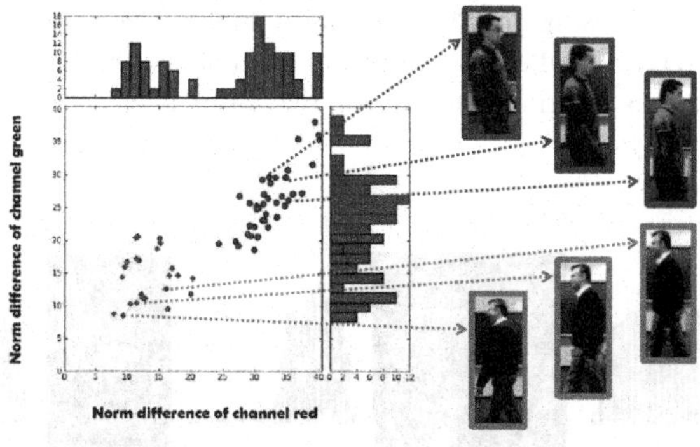

Fig. 4 Discrimination of the histogram of color space between blobs in taken from different frames

The tracking algorithm used is similar to other systems described in the literature. Once segmented foreground objects have been separated, we characterize the blob by its feature vector.

2.3 *Detecting Events and Behavior for Telecare*

For our initial work, we have considered a limited domain of events that we should detect, namely: arm gestures, and body positions upright or horizontal, to detect falls. These two cases are used to address anomalous behavior or simple help signals for elderly in their home environments. Our analysis is based upon comparing histogram moment distributions through the normalized difference of the histograms as well as the normalized difference of the moments of each histogram.

$$Hist(H_i,H_j) = \sum_{i>j} |H_i - H_j| \quad (4)$$

$$MHist(H_i,H_j) = \sum_{i>j} |M_i - M_j| \quad (5)$$

In order to test our histogram discriminatory technique, video events were recorded with very simple arm gestures, as shown in Figure 5. The foreground object was subtracted from the background by the methods previously described. For each of the histograms obtained in Figure 5, and for the histograms of Figure 6, statistical moments are calculated and then normalized histograms (normalized both by bins and total number of points) are obtained. Clustering can then be performed (similar to that of figure 4), by calculating $MHist(H_i,H_j)$, the normed difference. The histograms are obtained by summing all the pixels in the vertical direction.

The discrimination of the different body positions is possible comparing the moments of the histograms obtained (H_y - the vertical histogram). Figure 7 shows the results of different moments for frames shown in Figures 5 and 6. For example, the figure 5b (the central histogram) demonstrates a highly peaked third moment. For Figure 6, the difference in the distributions in the first and third moments is highly pronounced, and thus discrimination of the two cases is easily obtained from the simple moment analysis.

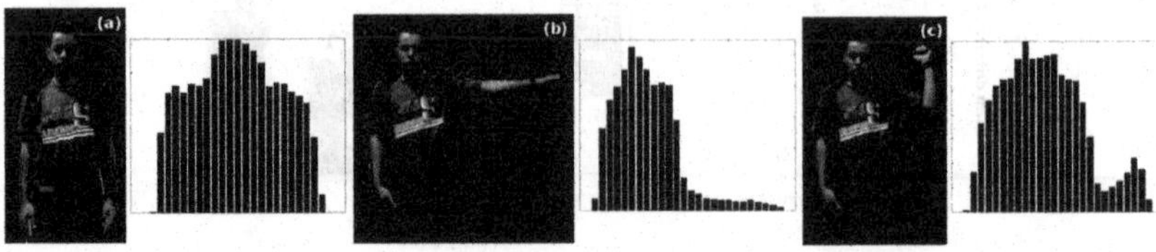

Fig. 5 Simple histogram results for obtaining moments for detecting arm gestures

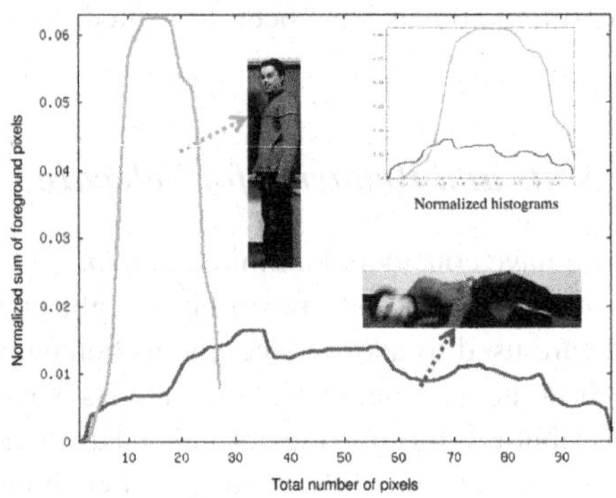

Fig. 6 Basic histogram technique used for discrimination body position. The inset image demonstrates the color space normalized to unity.

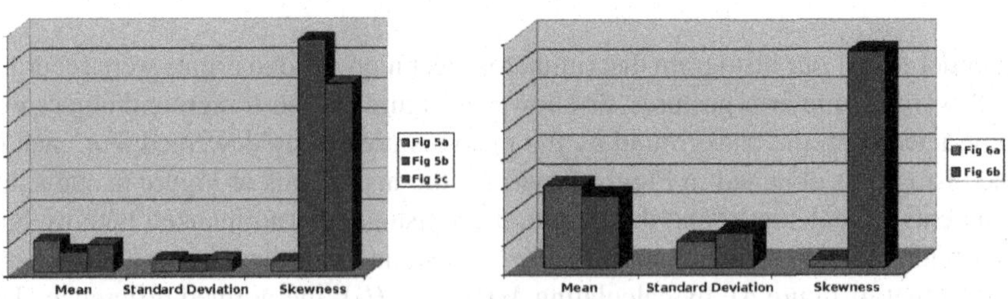

Fig. 7 Comparison of different histogram moments obtained from the video frames studied in Figure 5 and Figure 6

3 Experimental Results and Discussion

All the experimental tests and development were performed on a standard PC, consisting of an Intel Pentium D CPU 2.80GHz, with 2G of RAM and using the Ubuntu 9.10 Linux operating system. Videos and images were obtained from webcam with 2MPixel resolution.

The Figure 8 shows on a logarithmic scale the time results of the performance of the algorithms with a video of 30fps and 12 seconds of duration. The blue line represents the normal video reproduction, the magenta line is the video playing with our system without processing, the red color represents the foreground segmentation and the green line adds the time for processing blob clustering between each frame.

As shown in the previous section, the results of Figure 7 demonstrate that we can use statistical moment comparisons of histograms in order to discriminate between simple body positions. Thus, we have found that although our simple histogram techniques for human body position works well for some cases of interest and is easy to implement, it is not sufficiently robust. Because of its simplicity, however, we are presently improving the technique while at the same time investigating other algorithms.

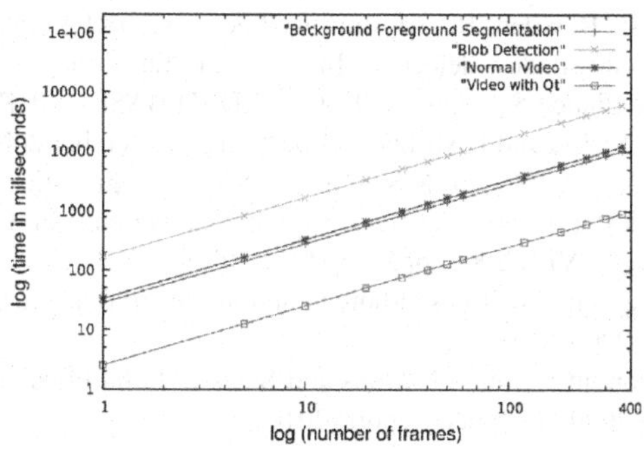

Fig. 8 Time of the different video reproductions

4 Conclusion

In this paper we have described preliminary work and algorithms on a software system which shall allow us to automatically track people and discriminate basic human motion events. This system is actually part of a more complete tele-monitoring system under development by our research group. The complete system shall include additional information from sensors, providing a complete information about a patient in their home. In this paper, however, we have restricted the study to video algorithms that shall allow us to identify body positions, in order to translate this information from a low level signal to a higher semantic level.

The paper provides encouraging result and opens many possibilities for future study. In particular, in the field of segmentation, the quantative comparison we described is an effective methodology which can be used to *optimize* parameters in each model. While the feature based tracking that used in this paper is rudimentary, a future study could combine this information with modern sequential Monte Carlo methods in order to obtain a more robust tracking. Finally, while the histogram model developed in this paper provides detection for a limited set of actions and events, it is a fast *real-time* method, that should have utility in real systems.

Acknowledgements. This work was supported by the Xunta de Galicia under the grant 08SIN002206PR.

References

1. Bradski, G., Kaehler, A.: Learning OpenCV: Computer Vision with the OpenCV Library. O'Reilly, Cambridge (2008)
2. Forsyth, D.A., Arikan, O., Ikemoto, L., O'Brien, J., Ramanan, D.: Computational studies of human motion: part 1, tracking and motion synthesis. Found. Trends. Comput. Graph. Vis. 1(2-3), 77–254 (2005)
3. Hu, W., Tan, T., Wang, L., Maybank, S.: A survey on visual surveillance of object motion and behaviors. IEEE Transactions on Systems, Man, and Cybernetics, Part C: Applications and Reviews 34(3), 334–352 (2004)
4. Kaewtrakulpong, P., Bowden, R.: An improved adaptive background mixture model for realtime tracking with shadow detection. In: Proc. 2nd European Workshop on Advanced Video Based Surveillance Systems, AVBS01, VIDEO BASED SURVEILLANCE SYSTEMS: Computer Vision and Distributed Processing (September 2001)
5. Meeds, E.W., Ross, D.A., Zemel, R.S., Roweis, S.T.: Learning stick-figure models using nonparametric bayesian priors over trees. In: IEEE Conference on Computer Vision and Pattern Recognition, CVPR 2008, pp. 1–8 (June 2008)
6. Poppe, R.: A survey on vision-based human action recognition. Image and Vision Computing 28(6), 976–990 (2010)
7. Department of Economic United Nations and Social Affairs Population Division. World population ageing 2009. Technical report (2010), `http://www.un.org/esa/population/publications/WPA2009/WPA2009-report.pdf`
8. Wei, Z., Bi, D., Gao, S., Xu, J.: Contour tracking based on online feature selection and dynamic neighbor region fast level set. In: Fifth International Conference on Image and Graphics, ICIG 2009, pp. 238–243 (September 2009)

Design and Modelling of the Nocturnal AAL Care System

J.C. Augusto, H. Zheng, M. Mulvenna, H. Wang, W. Carswell, and P. Jeffers

Abstract. We present the modelling of a monitoring system which provides night-time care by detecting situations of concern and therapeutic interventions as the core technological component within an Ambient Assisted Living project. The modelling of processes and interactions allows early detection of problems in the strategy to be implemented through simulation and verification.

Keywords: Intelligent Environments, AAL, safety critical.

1 Introduction

Dementia is a group of symptoms caused by specific brain disorder. The 2003 World Health Report Global Burden of Disease [1] estimated that dementia contributed to 11.2% of all years lived with disability among people aged 60 and over. In the UK, the Dementia 2010 report revealed that it is over 800,000 people sufferers a form of dementia and the annual cost of dementia is £23bn. It is estimated that the number of sufferers will pass the one million mark before 2025 [2]. In 2009, the National Dementia Strategy was published with the aim to support people with dementia and their carers to live well with dementia. Telecare and assistive technology spans a number of objectives in the national strategy [3].

Generally, people with dementia exhibit the symptoms of memory loss, problems using language, changes in personality, disorientation, problems doing usual daily activities and disruptive or inappropriate behaviour. Wandering and incontinence are the top two causes of institutionalisation. This paper presents our work

J.C. Augusto · H. Zheng · M. Mulvenna · H. Wang · W. Carswell
University of Ulster (UK)
e-mail: {jc.augusto,h.zheng,md.mulvenna,hy.wang,
w.carswell}@ulster.ac.uk

P. Jeffers
Fold Housing Association (UK)
e-mail: Paul.Jeffers@foldgroup.co.uk

P. Novais et al. (Eds.): Ambient Intelligence - Software and Applications, AISC 92, pp. 109–116.
springerlink.com

on the design and modelling of the Ambient Assisted Living (AAL) care system to support people with dementia at night time. We explain how tool-supported methodologies from Software Engineering can guide the development of the core monitoring and actuation process of a MAS system.

The reminder of this paper is organised as follows. In section 2, we introduce the NOCTURNAL project followed by a description of the design and modelling software package, SPIN, in section 3. NOCTURNAL model design, simulation and verification are detailed in section 4. The conclusion is presented in section 5.

2 The Nocturnal Project

AAL services and technologies are designed to help extend the time that older people can live at home by "increasing their autonomy and assisting them in carrying out activities of daily life" [4]. The services offered may include support for functional, activity, cognitive, intellectual and sensory-related activities; for example, providing alarms to detect dangerous situations that are a threat to the user's health and safety, continuously monitoring the health and well-being of the user and the use of interactive and virtual services to help support the user.

These technology-enriched services have evolved from relatively simple telecare services such as emergency fall alarm provision into more sophisticated telehealth services supporting people with long-term chronic health conditions such as Alzheimer's disease. Along with this evolution in the provision of services, there is a parallel development in the sophistication of the technology that underpins the AAL services and in the complexity and volume of the data that is harvested from the sensors that monitor the activity of the user in their home setting. Such data can include movement information, device usage information, medication compliance data and other rich data that can inform decisions for AAL services.

In the NOCTURNAL project, data representing activities of people with dementia is gathered and analysed in order to create behavioural profiles for them. The goal of the project is to develop a solution that supports older people with mild dementia in their homes, specifically during the hours of darkness. This is a relatively new area of research and was identified as a key area of need for care recipients with dementia after [5] found in their literature review of papers reporting on nighttime care of people with dementia that only 7% of papers addressed nighttime specific issues with a further 26% focusing on night and day activities together. It is also of interest because of the negative impact that lack of sleep and consequent anxiety causes for the informal carer in the home of the person with dementia. The support provided is also relatively unusual in that the AAL services are focused on identifying negative behaviours at nighttime such as restlessness and wakefulness [6] and then responding to these behaviours with interventions, e.g., guidance, that are designed to have a therapeutic impact. The design is for the AAL services to provide reassurance, aid and guidance for the general behaviour of the care recipient and to support a stable circadian rhythm. The types of therapeutic interventions include musical, visual and lighting based.

3 SPIN

SPIN [7] is one of the most well-known and used system for verification of software. It is a highly efficient and stable system with a user friendly interface, and good team support. SPIN is focused on the concept of models. Users can model a system by using a language called Promela (PROcess MEta-LAnguage), which emphasizes the role of processes and their interactions. Once a user has built a model the possible scenarios represented in the model can be simulated in various ways (e.g., randomly guided by the machine or user guided). Additionally users can perform what in Software Engineering is called Formal Verification, i.e., an exhaustive analysis of the computations implied by the model. SPIN provides a formal language for users to specify properties which can then be checked by the tool. If the property the user was trying to verify in the model does not hold then SPIN provides a counter-example (an explanation of why that property is not true for that particular model). The presentation below is more centred on the model and the simulation process (see [8, 9] for an emphasis on verification of IEs).

4 The Nocturnal Model

Our team is using SPIN to inform the design and modelling phases of the Nocturnal project. Some models consider the overall system whilst other models focus specific agents. The model of the overall architecture is depicted in Fig. 1.

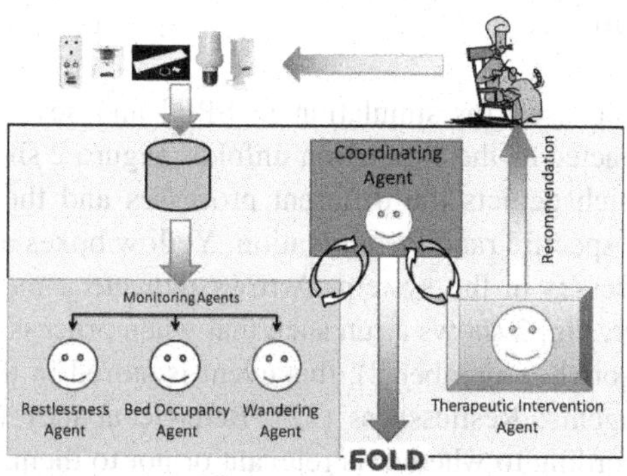

Fig. 1 Nocturnal Architecture

Activities of the client trigger sensors which are recorded as events in a database. These events are fed to a group of monitoring agents specialized on night related situations (e.g., restlessness, bed occupancy and wandering). When the number of episodes of interest detected by any single agent is above an acceptable threshold, which is dynamically adapted to the client and the context, the agent involved contacts a coordinating agent (CA) which have a holistic view of the context informed by all the single agent's reports. If appropriate, CA can trigger a therapeutic intervention with the aim of helping the client. If subsequent reports from the monitoring agents show there is still reasons for concern the coordinating agent can issue a new intervention or eventually if the situation requires it the call centre at Fold can be contacted so that a human deals directly with the situation. The system then is used as a way to increase independence with safety and to focus human interventions on those cases where is really needed.

4.1 Promela Model

Figure 1 depicted the main actors and interactions within the Nocturnal system architecture. The model was conceived to explicitly represent those elements and relationships to test the idea and use it as a framework to experiment with and discover non-trivial features which escaped the initial analysis of the team. One of the versions of the model of the overall architecture is provided in Appendix A (see comments inserted providing explanation of the model). Each main element of technology and human actors depicted in Figure 1 is represented in the Promela model by a process type (each of them has their name highlighted in boldface). Processes are autonomous entities running concurrently. Interaction amongst these elements is represented by message passing through synchronous channels. Naturally there are many features of the model that can be changed to experiment with, this model is only a snapshot in the lifetime of the system.

4.2 Simulation

This model can be used for simulation in SPIN and several different types of views can be extracted as the simulation unfolds. Figure 2 shows the Message Sequence Chart which depicts the different processes and the messages they sent each other in this specific random simulation. Yellow boxes at the top indicate the name of the processes in the system. Arrows indicate a message sent from one process to another. Fig. 2 shows a run such that when process Client activates (box number 6) a sensor (box number 7), this event is stored in the DB and passed to the monitoring agents: Restlessness (31), Bed Occupancy (10) and Wandering (37), they act according to whether is relevant or not to them. They in turn (in any possible combination, i.e., none, some or all) may or may not contact the Coordinating Agent (39). This one may (75) or many not (39) decide that a Therapeutic Intervention is needed. The therapeutic intervention may sometimes produce a response from the user (76-78). Sometimes the situation may require to contact Fold (199).

Notice this model does not focus on frequencies but rather on possibilities, i.e., whether something can be achieved or not within that architecture. Other modes we have explored focused on different aspects of the system, for example on how the monitoring agents can effectively keep track of the frequency of restlessness episodes, detect absence from bed or wandering, during a period of time.

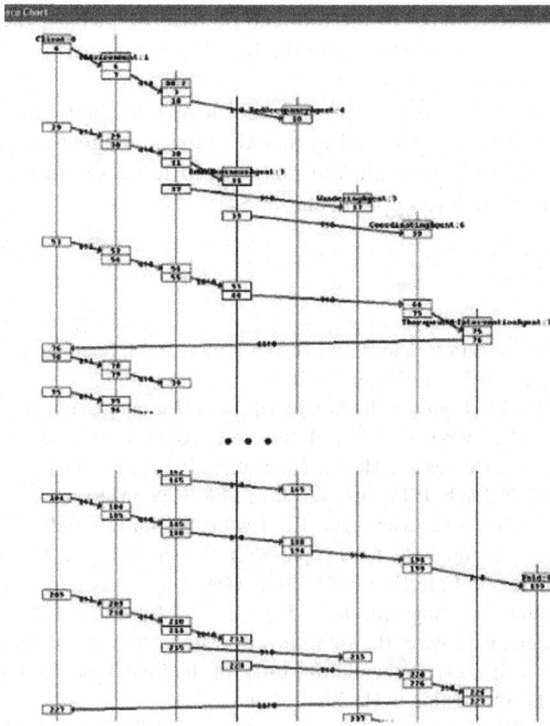

Fig. 2 Processes of the Overall Architecture communicate with each other

4.3 Verification

Formally specified behavioural properties (e.g., [] <> activeClient) can be explored using SPIN. These properties are usually related to the requirements of the system being examined. Examples of such properties for the Nocturnal case are: “can the system monitor the client continuously?”, “Are all sensor activations stored in the DB”, “Is each emergency followed by a therapeutic intervention?”.

5 Conclusions

The project Nocturnal aims at providing Ambient Assisted Living with an emphasis on night-time care. This paper has presented the underlying processes related to the project Nocturnal, and has shown how the use of well established techniques and tools from Software Engineering can be used to model these processes and to rigorously examine them during software development.

Our tool of choice provide simulation and verification capabilities. Through simulation different scenarios can be recreated either randomly by the tool or guided by the user. Once this has provided confidence to the team on the strategy being modelled other features of the tool can be used to check if the strategy represented in the model is consistent with the behavioural properties that are present in the model.

Our team has used this tool at two different levels: to model the overall architecture (the focus of this paper) and to model individual agents, e.g., restlessness agent of Fig. 1. This process allowed the team to focus on the strategy and the logic of the processes before implementation.

References

1. World Health Organization, The World Health Report 2003: Shaping the future (2003)
2. Alzheimer's Research Trust, Dementia (2010)
3. Department of Health, Living well with dementia: A National Dementia Strategy (2009)
4. Wojciechowski, M., Xiong, J.: A User Interface Level Context Model for Ambient Assisted Living. Smart Homes and Health Telematics, 105–112 (2008) (Online)
5. Carswell, W., McCullagh, P.J., Augusto, J.C., Martin, S., Mulvenna, M.D., Zheng, H., Wang, H.Y., Wallace, J.G., McSorley, K., Taylor, B., Jeffers, W.P.: A Review of the Role of Assistive Technology for People with Dementia in the Hours of Darkness. Technology and Health Care 17(4), 281–304 (2009)
6. Wang, H., Zheng, H., Augusto, J.C., Martin, S., Mulvenna, M.D., Carswell, W., Wallace, J., Jeffers, P., Taylor, B., McSorley, K.: Monitoring and Analysis of Sleep Pattern for People with Early Dementia. In: Proc. of the First Workshop on Know. Eng., Discovery and Dissemination in Health (2010)
7. Holzmann, G.J.: The SPIN Model Checker: Primer and Reference Manual. Addison, -Wesley (2003)
8. Augusto, J.C., McCullagh, P.: Ambient Intelligence: Concepts and Applications. Int. Journal on Computer Science and Information Systems 4(1), 1–28 (2007)
9. Augusto, J.C.: Increasing Reliability in the Development of Intelligent Environments. In: Pro-c. of 5th Int. Conf. on Intelligent Environments (IE 2009), Barcelona, Spain, pp. 134–141 (2009)

Appendix A: *Nocturnal Architecture Promela Model*

```
/* Data Structures Declarations Section                                    */
   bool activeSensor= true; bool contact= true; bool takeAction= true;
   bool intervention= true; bool event = true;
   /* all channels declared synchronous, i.e., handshake guaranteed  */
   chan client2sensors  = [0] of {bool};  /* Client to Sensors */
   chan env2DB = [0] of {bool}; /*  environment to DB */
   chan db2RA = [0] of {bool};  /* DB to Restlessness Agent */
   chan db2BOA = [0] of {bool};  /* DB to Bed Occupancy Agent */
   chan db2WA  = [0] of {bool};  /* DB to Wandering Agent */
   chan ra2CA  = [0] of {bool};  /* Restlessness Agent to Coordinating Agent */
   chan boa2CA  = [0] of {bool};  /* Occupancy Agent to Coordinating Agent */
   chan wa2CA  = [0] of {bool};  /* Wandering Agent to Coordinating Agent */
   chan ca2TIA  = [0] of {bool};  /* CA to Therapeutic InterventionAgent */
   chan tia2client  = [0] of {bool};  /* Therapeutic InterventionAgent to Client */
   chan ca2fold  = [0] of {bool};  /* Emergency Notification */
/*Process Declaration Section                                              */
   active proctype Client ()  /* represents free will of human */
   { bool activeC;
   end: do
         :: tia2client?intervention -->
            if  :: client2sensors!event
                :: skip   fi
                :: activeC --> atomic{client2sensors!event; activeC=false}
                :: !activeC --> activeC=true
         od }
   active proctype environment () /* generates sensor data and stores it in DB */
   { bool idle;
   end: do
         :: !idle -->atomic{client2sensors?event; env2DB!activeSensor; idle=true}
         :: idle --> idle=false
        od}
   active proctype DB () /* stores sensor data and passes it to agents*/
   { end: do
         :: env2DB?activeSensor -->
                atomic{
                    if :: db2RA!activeSensor
                        :: skip fi;
                    if :: db2BOA!activeSensor
                        :: skip fi;
                    if :: db2WA!activeSensor
                       :: skip fi              }
        od }
   active proctype RestlessnessAgent () /* detects restlessness episodes */
   { end: do
```

```
        :: db2RA?activeSensor -->
          if :: atomic{
                    printf("Restlesness Agent was interested on this information");
                    ra2CA!contact}
             :: skip  fi
        od }
active proctype BedOccupancyAgent () /* detects out of bed episodes */
{ end: do
        :: db2BOA?activeSensor -->
          if :: atomic{
                    printf("BedOcc. Agent was interested on this information");
                    boa2CA!contact}
             :: skip  fi
        od }
active proctype WanderingAgent () /* detects wandering episodes */
{ end: do
        :: db2WA?activeSensor -->
          if :: atomic{
                    printf("Wandering Agent was interested on this information");
                    wa2CA!contact}
             :: skip  fi
         od }
active proctype CoordinatingAgent ()  /* gathers advice from agents and,
when necessary, intervines in environment */
{ end: do
        :: ra2CA?contact -->
          if :: ca2TIA!takeAction
             :: ca2fold!takeAction
             :: skip fi
        :: boa2CA?contact -->
          if :: ca2TIA!takeAction
             :: ca2fold!takeAction
             :: skip fi
        :: wa2CA?contact -->
          if :: ca2TIA!takeAction
             :: ca2fold!takeAction
             :: skip fi
         od}
active proctype TherapeuticInterventionAgent () /* actuates in env. to
 achieve goals set by coordinator agent */
{ end: do :: ca2TIA?takeAction -->
                      atomic{tia2client!intervention; printf("Action Taken!")} od }
active proctype Fold () /* deals with emergencies */
{ end: do :: ca2fold?takeAction --> printf("Action Considered by Fold!") od }
```

The EducAgent Platform: Intelligent Conversational Agents for E-Learning Applications

David Griol, Jesús García-Herrero, and José M. Molina

Abstract. In this paper, we describe a multi-agent system developed for teaching support and student's self-learning. The main objective of the *EducAgent* platform is the creation of an innovative virtual space following the principles of the European Higher Education Area to make subjects and e-learning initiatives to become a more flexible, participatory and attractive space. One of the most important characteristics of the developed platform is to facilitate a more natural interaction between the system and students by means of conversational agents. We describe the main features of the *EducAgent* platform and its application in the new European Computer Science Degree at the Carlos III University of Madrid.

Keywords: Conversational Agents, E-Learning, Oral Interaction, Intelligent Agents, Education and New Technologies.

1 Introduction

Ambient Intelligence (AmI) emphasizes on greater user-friendliness, more efficient services support, user-empowerment, and support for human interactions. In this vision, people will be surrounded by intelligent and intuitive interfaces embedded in everyday objects around us and an environment recognizing and responding to the presence of individuals in an invisible way [1]. To ensure such a natural and intelligent interaction, it is necessary to provide an effective, easy, safe and transparent interaction between the user and the system. This way, conversational agents [6], which marry agent capabilities with computational linguistics, have became a strong alternative to enhance multi-agent systems with intelligent communicative capabilities, as speech is one the most natural and flexible means of communication among humans.

David Griol · Jesús García-Herrero · José M. Molina
Group of Applied Artificial Intelligence (GIAA), Computer Science Department,
Carlos III University of Madrid
e-mail: {david.griol,jesus.garcia,josemanuel.molina}@uc3m.es

P. Novais et al. (Eds.): Ambient Intelligence - Software and Applications, AISC 92, pp. 117–124.
springerlink.com

With the growing maturity of conversational technologies, the possibilities for integrating conversation and discourse in e-learning are receiving greater attention, including tutoring [7], question-answering [9], conversation practice for language learners [3], pedagogical agents and learning companions [2], and dialogs to promote reflection and metacognitive skills [5].

In addition, learning and training institutions are growingly facing important challenges. In fact, the new framework of the European Higher Education Area (EHEA) involves teaching mainly oriented to the attainment of competencies, so that the role of the teacher is to facilitate and guide students to intellectually access contents and professional practices corresponding to their degrees. To achieve this objective, more participatory and reflective teaching methodologies are required to allow students reaching the maximum autonomous academic and personal development as possible. In this space, teachers are not only transmitters of knowledge, but also become professionals who create and organize complex learning environments, involving students in their own learning process by means of appropriate strategies.

In this framework, according to Roda et al. [8], enhanced e-learning systems are expected to i) accelerate the learning process, ii) facilitate access, iii) personalize the learning process, and iv) supply a richer learning environment. To do this, and promoted by the introduction of the new European degrees at Carlos III University of Madrid, continuous assessment has been introduced based on students' effort and active participation in their learning. This way, students are followed up through activities which promote their participation and knowledge acquisition, such as initiatives which enable them to know their progresses.

Following these premises to provide enhanced e-learning initiatives in the EHEA framework, we have developed the *EducAgent* platform. The main objective is to develop a flexible learning tool to offer the stimulus, support and environment necessary to guide students in a continuous and active learning process. To do this, intelligent conversational agents have been included in the architecture of our platform, which facilitate a more natural interaction between the system and students, select the different contents, present them to students, collect their answers, and provide a feedback after the analysis.

2 The *EducAgent* Platform

The *EducAgent* platform consists of a set of intelligent agents which works as a virtual space where students can interact to be presented cases and problems to be solved, which are adapted to their progresses and takes into account specific problems detected during the student's evolution during the course. In addition, student's interaction with the platform also allow to automatically assess their learning. To do this, we bring to our teaching the most recent advances made by the members of the project in different research fields.

Three main features are highlighted to be present in most of the educational intelligent agents. Firstly, the ability of *communication*. Intelligent agents can communicate with the user, other agents and other programs. The user communicates with a user-friendly interface to customize their preferences. For the development of the *EducAgent* platform, we have defined communication with students in the most natural way possible as a priority. For this reason, the communication in our platform is carried out by means of advanced conversational agents. To successfully manage the interaction with users, conversational agents are usually developed following a modular architecture, which generally includes the following tasks: automatic speech recognition (ASR), natural language understanding (NLU), dialog management (DM), database management (DB), natural language generation (NLG) and text-to-speech synthesis (TTS). The development of these agents implies the achievement of a set of challenges, which depend on the selected languages models, dialog initiatives, confirmation strategies, responses generation, etc. Our proposal is the incorporation of statistical methodologies to deal with all these important design decisions [4].

The second main feature is the degree of *intelligence*, which has a wide range of possibilities often achieved by means of the incorporation of technologies from Artificial Intelligence. In our case, Speech Technologies and Natural Language Processing are used to facilitate the automatic analysis of students' responses. Finally, the third main feature is *autonomy*. An agent must not only be able to make suggestions, but also to act with proactiveness. In *EducAgent*, agents are implemented with the autonomy required to select which are the most appropriate contents to be presented to students and which is the most appropriate response that must be given as a result of the analysis of the interaction. Different natural language e-learning applications also offering learning reinforcement to review self-assessments are detailed in [5].

2.1 Architecture of the EducAgent Platform

Figure 1 shows the architecture designed for the EducAgent platform to select the different contents and generate a questionnaire, perform the interaction with students by means of a conversational agent, carry out the corresponding analysis of the students' answers, and provide them with the appropriate feedback.

As Figure 1 shows, for each one the units that make up the course, the platform selects from a database the different questions corresponding to the concepts selected for the evaluation and presents. The generated questionnaire is presented to the student using the Moodle platform and by means of the Opera browser. This way, students access the questionnaire on the web, and they can answer the questions using their voice, keyboard and/or mouse. To do so, these questionnaires are developed using the VoiceXML standard[1].

To allow students answering the different questions using their voice, the platform uses the ViaVoice speech recognition technology from IBM, embedded in the Opera browser. Once the text transcription of the speech has been obtained, the Language Understanding Module generates the semantic interpretation of the input

[1] http://www.w3.org/TR/voicexml20/

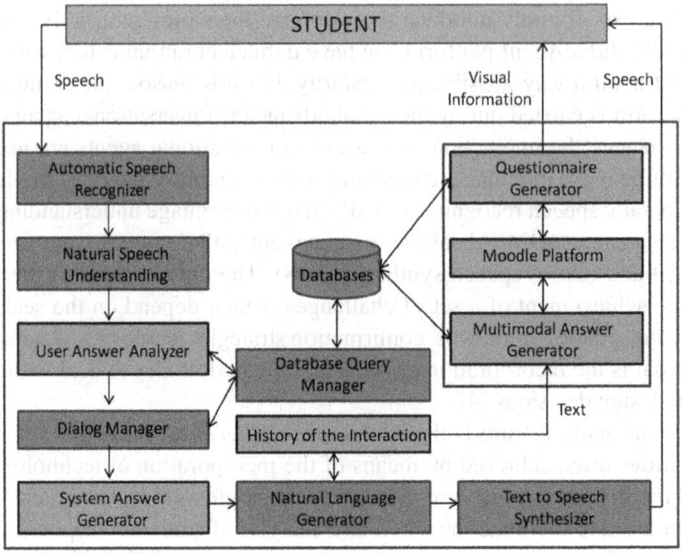

Fig. 1 Architecture of the EducAgent platform

sentence. To do this, this module analyzes the contents provided by the students by means of specific grammars generated for each one of the questions. By obtaining the meaning of the student's utterance and using the result of its comparison with the correct answer defined by the teacher (contained in the database system and accessible by means of the database query module), the User Answer Analyzer calculates the percentage of success and the set of recommendations to be made to the student. This is done by means of grammars in which the student's answer is compared with the reference answer, assigning a specific score and answer each time a coincidence is detected. This way, in our system we provide a balance between accuracy and flexibility in the evaluation process. Test questions provide total reliability on the correction of these answers, while our grammar-based functionality offers the flexibility of natural language.

Then, the Dialog Manager decides the next action of the platform, taking into account the analysis previously described for the utter utterance (for instance, to confirm information supplied by the student, to request additional information, to continue with the next question, etc.). The System Answer Generator takes into account the result of the analysis carried out for the previous set of system questions and generates the corresponding system answer, which is presented to the student by means of the result generated by the Natural Language Generation module. The generated text is formatted into VoiceXML code, which is presented using the Moodle platform, and by means of a synthesized voice that generates the oral answer corresponding to the generated answer.

This way, students can carry out several interactions with the platform, consulting contents corresponding to the different stages defined in the course, from which their can be evaluate their knowledge and extract conclusions about the results which must be obtained at the end of each of them. The students' interactions with the virtual agents provide essential information for both teachers and students. Teachers are provided with a feedback about the degree of the student's understanding of the different contents. The interaction with the platform allows students to develop the ability to put concepts into practice, verifying whether their proposed solution is correct or not and also doing this in an innovative environment.

3 Practical Case Study

The *EducAgent* platform started to be implemented at the end of the academic year 2009-2010 for its application in the Computer Science Degree at the Carlos III University of Madrid. The subject Compilers was selected for the elaboration of the different questionnaires for the platform. The methodology defined for the evaluation of this subject emphasizes students' continuous assessment. For the generation of the different questions and practical cases to be included i the platform, we considered the following types of exercises: i) questions concerning theoretical contents as a review of methodologies and concepts; ii) connection with programs (like Flex[2] and Yacc[3]) used to propose practical implementations and provide code execution; iii) practical cases proposed to the student to obtain conclusions about the appropriate processes for resolve specific problems.

A total of 110 questions, practical cases and problems was elaborated for the subject. We evaluate these questions by proposing them to students. It should be emphasized that the total of questions were answered by 89% of students. A 86% percentage of students expressed the usefulness of the provided cases and problems to facilitate the achievement of the objectives of the course, enhance their learning and facilitate knowing the degree of understanding of the different contents.

Figure 2 shows an example of a VoiceXML form for the multimodal presentation of a specific question, collect the student's response and provide it to the Natural Language Understanding Module in the *EducAgent* platform. As it can be seen, each VoiceXML file corresponding to a specific question includes an initial grammar, manages of help events, and deliveries the student's answer to the Language Understanding and Answer Analyzer modules.

Figure 3 shows the composition of a questionnaire with questions related to Lexical Analysis. It can be seen that the different kinds of previously described exercises have been incorporated. For each one of the proposed questions, the system provides a comparison of the student's responses with the solution proposed by the agent, as well as instructions about possible mistakes, important points which must be enhanced, information about typical errors, etc.

[2] http://flex.sourceforge.net/

[3] http://sourceforge.net/projects/byacc/

```
<?xml version="1.0" encoding="UTF-8"?>
<vxml version="2.0" xmlns="http://www.w3.org/2001/vxml"
xmlns:xsi="http://www.w3.org/2001/XMLSchema-instance"
xsi:schemaLocation="http://www.w3.org/2001/vxml
http://www.w3.org/TR/voicexml20/vxml.xsd">
<form id="Lexical Analysis">
<block>Please, answer now the following question.</block>
 <field name="PAL-03">
  <prompt>Do you think that regular expressions can be used
       to recognize tokens? Why?</prompt>
  <grammar src="pal-03.grxml"  type="application/srgs+xml"/>
  <catch event="help">
     You have to explain if it is possible to recognize tokens by means
     of the definition of regular expressions.<reprompt/>
  </catch>
</field>
<submit next="/servlet/analisis/pal" namelist="pal-03"/>
</block>
</form>
</vxml>
```

Fig. 2 Example of a VoiceXML form to generate a specific question in *EducAgent*

The indicators about the operation of the platform that we want obtain at the end of the current academic year include the complete evaluation of the different agents and educational contents for the subject, the evaluation of their reliability and usability, the validation of the acceptation degree of the different contents, and the definition and evaluation of technological parameters associated to the specific operation of each one of the modules in the platform.

We have already completed a preliminary evaluation of the *EducAgent* platform based on questionnaire to assess the students' subjective opinion about the system performance. The questionnaire had 10 questions: i) Q1: *State on a scale from 1 to 5 your previous knowledge about new technologies for information access.*; ii) Q2: *How many times have you used Opera Voice before?*; iii) Q3: *How well did the system understand you?*; iv) Q4: *How well did you understand the messages generated by the system?*; v) Q5: *Was it easy for you to get the requested information?*; vi) Q6: *Was the interaction rate adequate?*; vii) Q7: *Was it easy for you to correct the system errors?*; viii) Q8: *Were you sure about what to say to the system at every moment?*; ix) Q9: *Do you believe the system behaved similarly as a human would do?*; x) Q10: *In general terms, are you satisfied with the* EducAgent *platform?*.

The possible answers for each one of the questions were the same: *Never*, *Seldom*, *Sometimes*, *Usually*, and *Always*. All the answers were assigned a numeric value between one and five (in the same order as they appear in the questionnaire). Table 1 shows the average, minimal and maximum values for the subjective evaluation carried out by a total of 15 students from one of the groups in the subject.

From the results of the evaluation, it can be observed that students positively evaluates the facility of obtaining the data required to fulfill the complete set of objectives of the proposed in the exercises defined for the subject, the suitability

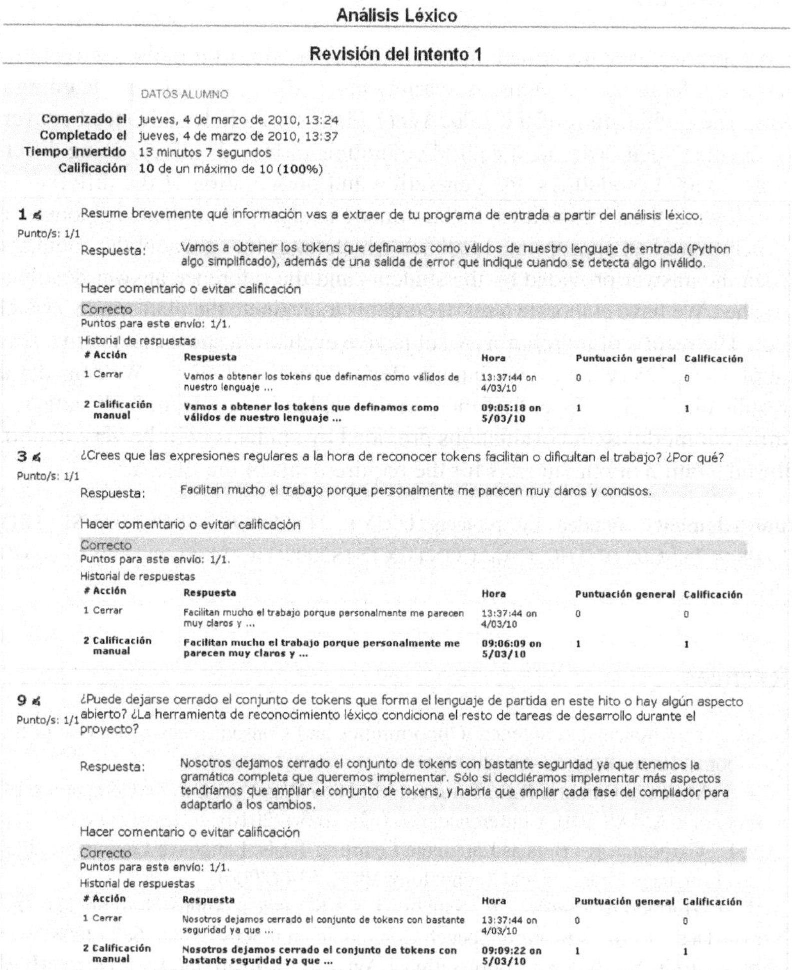

Análisis Léxico

Revisión del intento 1

DATOS ALUMNO

Comenzado el jueves, 4 de marzo de 2010, 13:24
Completado el jueves, 4 de marzo de 2010, 13:37
Tiempo invertido 13 minutos 7 segundos
Calificación 10 de un máximo de 10 (**100%**)

1 Resume brevemente qué información vas a extraer de tu programa de entrada a partir del análisis léxico.
Punto/s: 1/1

Respuesta: Vamos a obtener los tokens que definamos como válidos de nuestro lenguaje de entrada (Python algo simplificado), además de una salida de error que indique cuando se detecta algo inválido.

Hacer comentario o evitar calificación

Correcto
Puntos para este envío: 1/1.
Historial de respuestas

# Acción	Respuesta	Hora	Puntuación general	Calificación
1 Cerrar	Vamos a obtener los tokens que definamos como válidos de nuestro lenguaje ...	13:37:44 on 4/03/10	0	0
2 Calificación manual	**Vamos a obtener los tokens que definamos como válidos de nuestro lenguaje ...**	**09:05:18 on 5/03/10**	**1**	**1**

3 ¿Crees que las expresiones regulares a la hora de reconocer tokens facilitan o dificultan el trabajo? ¿Por qué?
Punto/s: 1/1

Respuesta: Facilitan mucho el trabajo porque personalmente me parecen muy claros y concisos.

Hacer comentario o evitar calificación

Correcto
Puntos para este envío: 1/1.
Historial de respuestas

# Acción	Respuesta	Hora	Puntuación general	Calificación
1 Cerrar	Facilitan mucho el trabajo porque personalmente me parecen muy claros y ...	13:37:44 on 4/03/10	0	0
2 Calificación manual	**Facilitan mucho el trabajo porque personalmente me parecen muy claros y ...**	**09:06:09 on 5/03/10**	**1**	**1**

9 ¿Puede dejarse cerrado el conjunto de tokens que forma el lenguaje de partida en este hito o hay algún aspecto abierto? ¿La herramienta de reconocimiento léxico condiciona el resto de tareas de desarrollo durante el proyecto?
Punto/s: 1/1

Respuesta: Nosotros dejamos cerrado el conjunto de tokens con bastante seguridad ya que tenemos la gramática completa que queremos implementar. Sólo si decidiéramos implementar más aspectos tendríamos que ampliar el conjunto de tokens, y habría que ampliar cada fase del compilador para adaptarlo a los cambios.

Hacer comentario o evitar calificación

Correcto
Puntos para este envío: 1/1.
Historial de respuestas

# Acción	Respuesta	Hora	Puntuación general	Calificación
1 Cerrar	Nosotros dejamos cerrado el conjunto de tokens con bastante seguridad ya que ...	13:37:44 on 4/03/10	0	0
2 Calificación manual	**Nosotros dejamos cerrado el conjunto de tokens con bastante seguridad ya que ...**	**09:09:22 on 5/03/10**	**1**	**1**

Fig. 3 Example of a questionnaire created using the *EducAgent* platform (in Spanish)

Table 1 Results of the evaluation of the *EducAgent* platform (1=worst, 5=best evaluation)

	Q1	Q2	Q3	Q4	Q5	Q6	Q7	Q8	Q9	Q10
Average Value	4.6	2.8	3.6	3.8	3.2	3.1	2.7	2.3	2.4	3.3
Maximum Value	5	3	4	5	5	4	3	3	4	4
Minimal value	4	1	2	3	2	3	2	2	1	3

of the interaction rate during the dialog. The sets of points that they mention to be improved include the correction of system errors and a better clarification of the set of actions expected by the platform at each time.

4 Conclusions

In this paper we have described the main characteristics of a multi-agent platform developed to facilitate autonomous learning and self-assessment for e-learning initiatives. The architecture of the *EducAgent* platform includes advanced conversational agents which facilitate a natural communication with student using different input and output modalities, the generation and presentation of the different contents following these modalities, the automatic analysis of students responses, and the generation of an appropriate feedback which takes into account the comparison between the answer provided by the students and the reference answer detailed by the teacher. We have elaborate a set of contents to evaluate the platform in a specific subject. The results of a preliminary subjective evaluation show the positive acceptation of a set of very important features defined for the platform. We consider that the results obtained at the end of the current academic year from both statistics of the different modules and evaluations provided by students, will be very important for the adaptation of our subjects for the requirements of the EHEA.

Acknowledgments. Funded by projects CICYT TIN2008-06742-C02-02/TSI, CICYT TEC2008-06732-C02-02/TEC, CAM CONTEXTS (S2009/TIC-1485), and DPS2008-07029-C02-02.

References

1. Augusto, J.: Ambient Intelligence: Opportunities and Consequences of its Use in Smart Classrooms. Italics 8(2), 53–63 (2009)
2. Cavazza, M., de la Camara, R.S., Turunen, M.: How Was Your Day? a Companion ECA. In: Proc. of AAMAS 2010 Conference, pp. 1629–1630 (2010)
3. Fryer, L., Carpenter, R.: Bots as Language Learning Tools. Language Learning and Technology. Language Learning and Technology **10**(3), 814 (2006)
4. Griol, D., Hurtado, L., Segarra, E., Sanchis, E.: A Statistical Approach to Spoken Dialog Systems Design and Evaluation. Speech Communication 50(89), 666–682 (2008)
5. Kerly, A., Ellis, R., Bull, S.: Conversational Agents in E-Learning. In: Proc. of AI-2008, pp. 169–182 (2008)
6. McTear, M.F.: Spoken Dialogue Technology: Towards the Conversational User Interface. Springer, Heidelberg (2004)
7. Pon-Barry, H., Schultz, K., Bratt, E.O., Clark, B., Peters, S.: Responding to student uncertainty in spoken tutorial dialogue systems. International Journal of Artificial Intelligence in Education 16, 171–194 (2006)
8. Roda, C., Angehrn, A., Nabeth, T.: Conversational Agents for Advanced Learning: Applications and Research. In: Proc. of BotShow 2001 Conference, p. 17 (2001)
9. Wang, Y., Wang, W., Huang, C.: Enhanced Semantic Question Answering System for e- Learning Environment. In: Proc of AINAW 2007 Conference, pp. 1023–1028 (2007)

Improving Human Face Detection through TOF Cameras for Ambient Intelligence Applications

J.R. Ruiz-Sarmiento, C. Galindo, and J. Gonzalez

Abstract. One of the cornerstones of ambient intelligence technology is the need of sensory systems to reliably notice the presence of people. Several approaches for detecting humans within a non-controlled scenario can be found in the literature but they exhibit not enough effectiveness, i.e. a high rate of false positive or true negative detections. This becomes a drawback for the development of a variety of ambient intelligence applications which depend on such a sensory capability.

In this paper we propose the use of a TOF camera for noticing human presence by detecting their faces. Apart from a typical intensity image, this camera also provides a range image of the scene. The proposed methodology first detects faces from the intensity image (by using the Viola-Jones algorithm) and then analyzes those detections in the range image to discard false positives. Experimental evaluations of the proposed process have yielded excellent results in non-controlled scenarios, eliminating most of the false positive detections.

Keywords: Human Detection, Range (Time Of Flight) Cameras, Mobile Robotics.

1 Introduction

Ambient Intelligence (AmI) has recently come out to increase the comfort of people within their daily life. AmI refers to the enhancement of the environment through a set of devices that intelligently controls certain functionalities, pursuing a variety of aims. For instance, an intelligent fridge can take account on the aliments that it contains, automatically ordering those which have been already consumed. Such an intelligent fridge, equipped with gas sensors, could also alert the user when a particular meal is rotten [7]. Thus, the inhabitants of the home can be easygoing with their food supply, simplifying this chore. Other practical cases where AmI may help is in

J.R. Ruiz-Sarmiento · C. Galindo · J. Gonzalez
System Engineering and Automation Dpt., University of Malaga, Spain
e-mail: jotaraul@isa.uma.es, {cipriano, jgonzalez}@ctima.uma.es

P. Novais et al. (Eds.): Ambient Intelligence - Software and Applications, AISC 92, pp. 125–132.
springerlink.com

the adjustment of resources consumption at the same time that all the comforts of the environment are kept. A sensory system can detect the presence/absence of people in a given area to control, for example a heating device. Even more interesting is the possibility to estimate the number of people and to be able to recognize them in order to accordingly adjust the temperature of a room following pre-specified preferences. These are some examples that reveal the human-centric characteristic of Ambient Intelligence technology. Given that humans are the main actors, it becomes clear the interest of AmI in reliably noticing the presence of people in order to act adequately.

Computer vision contributes to this issue by providing algorithms to detect human faces in images [2, 11, 12], but their reliability are still not high enough for many real AmI applications. Very recently, a new type of sensor, called *TOF camera* (Time-Of-Flight camera), has appeared in the market providing both intensity and range data. These cameras permit us to improve the detection of the truly presence of human faces not only through visual information, e.g. skin-like colour, 2D shape, etc., but also by exploiting 3D characteristic information of human faces, for example, the area of the nose is prominent with respect to the area of the cheeks.

In the recent literature some works can be found exploring the capabilities of TOF cameras. Many of them address the physical characterization of TOF cameras [1, 13], and others focus on their application in a variety of fields [4, 14]. Some related works to ours are [3, 6], which propose simple and direct approaches for face detection using TOF cameras. In this paper we improve their achievements through a battery of 3D-shape tests that leads to a reliable and robust face detector. Our experiments have been successfully conducted on real and non-modified scenarios, yielding results with a high rate of true positives and reducing the number of false positives, which are a serious problem for known intensity-based face detectors. Some of our experiments have been conducted with the TOF camera mounted on a mobile robot, to prove the suitability of our approach for AmI applications that may incorporate a service robot.

Next we describe the proposed method and the evaluation results of the experiences conducted on real, non-controlled scenarios.

2 The Proposed Method

Our aim is to develop a robust face detection process using a TOF camera that produces the minimum number of false positives as possible, while maintaining high detection rates. Our approach is divided into two steps (see figure 1): the first one applies the well-known Viola-Jones classifier [11] over the intensity image to determine a set of candidates regions that presumably contain a human face. In the second step, the resulting candidates are checked against a battery of 3D-shape tests that operates over the range image. This second phase aims at discarding false positives of the first stage.

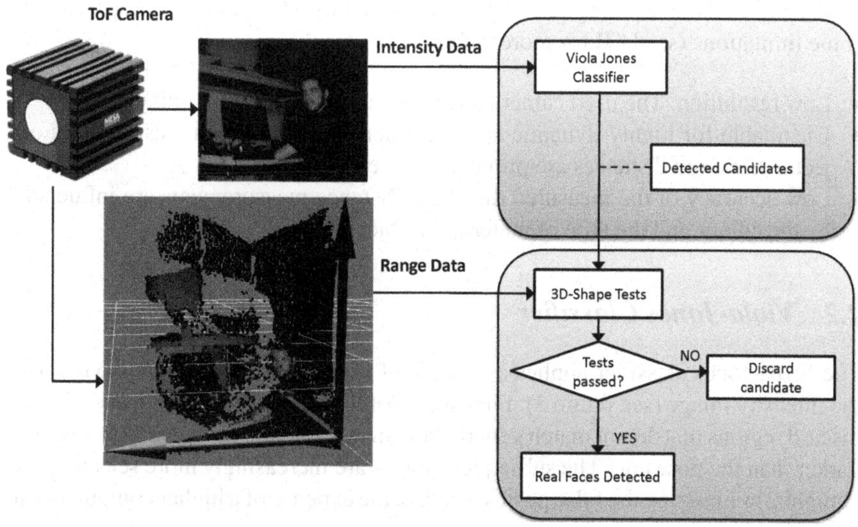

Fig. 1 Scheme of the proposed two-step process for face detection

2.1 TOF Cameras

Broadly speaking, a time-of-flight camera is a device that provides both intensity and range data, i.e. distance information to the sensed objects. In particular, the TOF camera considered in this work, manufactured by *Mesa Imaging* [8], performs by emitting a continuous wave modulation through an infrared led array (see figure 2). When the wave is reflected to the camera, their CCD/CMOS cells are excited with a signal that exhibits a certain phase shift, which permits the device to calculate the distance and reflectivity of the target through cross correlation up to a distance of 5 m. with an accuracy of ± 10 *mm*.

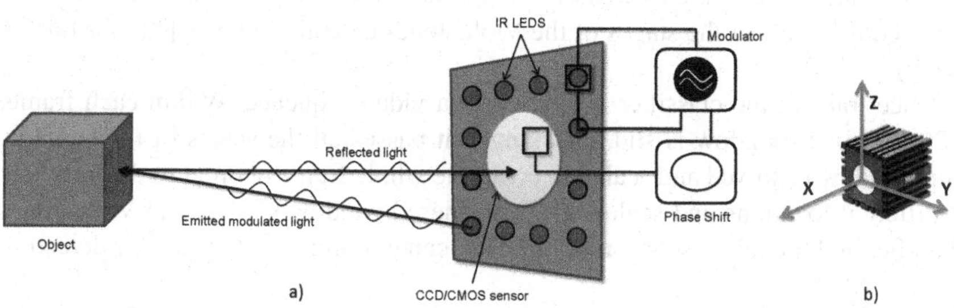

Fig. 2 a) Outline of the TOF camera operation. b) Reference system established by the used TOF camera.

In spite of the interesting possibilities provided by TOF cameras, they have still some limitations (see [13] for more details):

- Low resolution. The used camera has a resolution of 176 *x* 144 pixels.
- Unsuitable for highly dynamic scenes. When a scene presents fast moving objects, measured distances are prone to large errors.
- Low accuracy of the measured distance. Distance measurements are influenced by the colour and the type of material of objects.

2.2 Viola-Jones Classifier

The Viola-Jones classifier applies a cascade of tests of increasing complexity, over the intensity image (see figure 3). First stages of the cascade are simple and quickly discard regions that do not match general human faces' features, e.g. the eyes' area is darker than the nose one. The subsequent stages are increasingly more selective and complex, minimizing the false positive rate, at the expense of a higher computational burden.

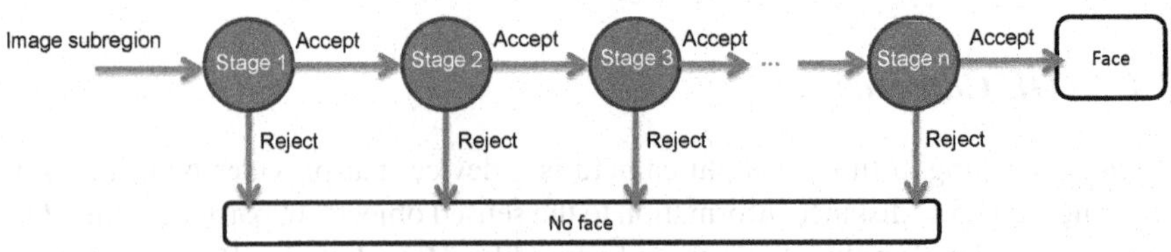

Fig. 3 Structure of a face detector cascade. Computationally less demanding stages are applied first.

Tests considered in each stage are trained using AdaBoost[15]. AdaBoost is a learning algorithm that chooses weak classifiers between a family of simple (*Haar*) features. These classifiers are combined into strong classifiers, which become constituents of the stages of the Viola-Jones cascade (refer to [11] for further detail).

Once trained, the classifier is applied to a video sequence. Within each frame, a 20*x*20 pixels window is slid, checking if it passes all the stages of the cascade. This window is moved and scaled to cover the whole frame in order to detect a face at different locations and scales. The considered window size for the Viola-Jones classifier and the low resolution of our TOF camera limit the distance for detecting faces to 2.5 meters.

Although this face detector provides good results when compared to other methods [10], the number of false positives it produces is a serious drawback for many AmI applications.

2.3 3D-Shape Tests

To overcome the problem of such a high number of false positives, we propose a second step where three 3D-shape tests are applied over the range image to confirm or not every candidate. Each test checks out if a particular 3D-shape feature is present in the candidate region or not, identifying and eliminating different types of false positives. Figure 4 shows some examples of false positives detected by the two of the proposed tests which are described in the following sections.

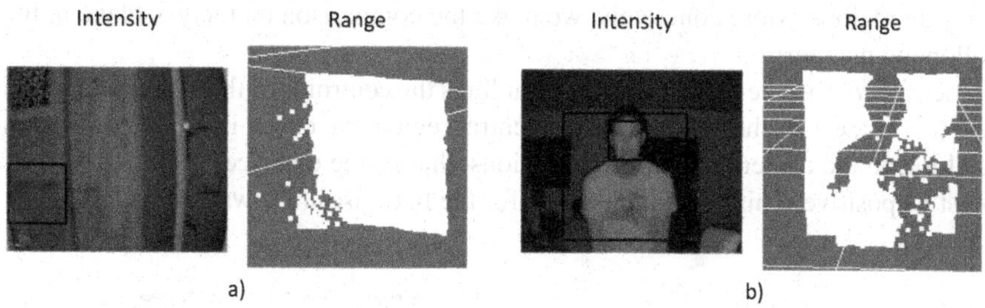

Fig. 4 a) A false positive detected because it is a flat region. b) A false positive because of an abnormal face size for that distance.

2.3.1 Test #1: Flat Region

The first test is based on the fact that human faces are not flat, so they present certain relief on the range image. An example of this type of false positives is shown in figure 4a. This test is implemented by computing the covariance matrix, C, of the spatial position (x,y,z) of the pixels that conform the candidate region. Eigenvalues of the C matrix give information about the spatial distribution of the pixels that form the region. Concretely, the lower the smallest eigenvalue, the flatter the region. Notice that a similar reasoning could be stated for the standard deviation of the x coordinates (distances to the camera) which is cheaper to compute, however, we have verified that it presents some problems for slightly oblique regions.

2.3.2 Test #2: Size-Distance Ratio

This test discards candidates that do not match the expected size of a human face at a given distance. Figure 4b is an example of a false positive detected by this test. In our implementation we have considered that the normal size of human faces, on average, is 290 cm^2 with a typical deviation of 60 cm^2 approximately. These figures have been obtained by analyzing around 10000 face images.

2.3.3 Test #3: Facial Structures

This test exploits the fact that human faces have a general common morphology. It segments the candidates using a growing regions method over the range data to extract the candidate face from the background, and then it divides it into nine subregions as shown in figure 5a. The depths of these subregions to the camera must verify certain constrains characteristic of human faces. For instance, the region that presumably contains the nose (numbered as 5 in figure 5a) should stand out with respect to the lateral subregions, that correspond to the cheeks, i.e. 4 and 6. An example of a false positive declared by this test is shown in figure 5b.

In the implementation of this test, the position of each region is set as the centroid of the pixels in it. Five comparisons are performed to check whether the central regions of the three rows and of the two diagonals are closer to the camera than the side regions. More concretely, we make the comparison on the y-x plane in the following manner:

Let $x = a \cdot y + b$ be the straight line that links the centroids of the side regions, and let $P_c = (y_c, x_c)$ be the centroid of the central region, we check if the central region is closer to the camera than the side regions, that is, the distance $d = a \cdot y_c + b - x_c$ must be positive. This checking is done for the five cases (3 rows + 2 diagonals).

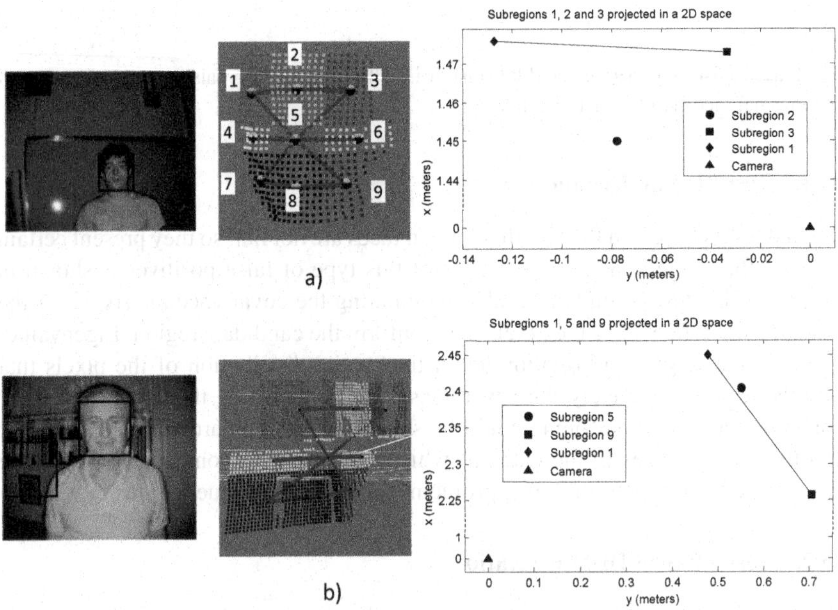

Fig. 5 a) Left: A candidate region. Middle: The centroids are represented with spheres while lines represent the constrain to be checked. Right: projection into the y-x plane. b) Left: A false positive (behind the person). Middle: Subdivision of the region and the restrictions to be fulfilled. Right: projection of three subregions into the y-x plane. Note that the test is not passed in this case.

3 Evaluation

In order to prove the effectiveness of our approach we have conducted 16 experiments in two scenarios: *fixed-camera*, where the TOF camera was fixed while people were moving around, and *moving-camera*, where the camera was mounted on a mobile robot, and thus, it was affected by important changes in illumination. An example video of an AmI application where a mobile robot detects people within their working place can be watched at `http://www.youtube.com/watch?v=GJE4A7R6LNs`.

We apply the Viola-Jones cascade classifiers from the OpenCV library, concretely the one labeled as *haarcascade_frontalface_alt2*, for detecting candidate regions. Although it provides good results [9], a considerable rate of false positives are produced. In our experiments from a total of 11.184 candidates, 685(5.77%) was the number of false positives. These false positives were identified by visual inspection.

The proposed 3D-shape tests have been implemented in a multi-thread C++ program, using a Intel®Core™2 Quad CPU Q6600 2.4GHz, with 4Gb RAM, which permits us to process up to 14 fps. We take advantage of the multi-core capability to run simultaneously all the tests, albeit a sequential execution solution could be also adopted.

When the 3D-shape tests are considered the number of false positives is drastically reduced, as shown in table 1. Notice that each test by itself has a modest ratio of false positive declaration, but when combined they yield excellent results, eliminating all the false positives in the fixed-camera scenario and around the 98% of cases with the camera onboard the robot. Regarding the false negative rate, i.e. faces which are erroneously neglected, the battery of the three tests discards around 3% of actual faces in both scenarios, which does not represent a serious inconvenience since we are dealing with a video sequence.

Table 1 False positives detected by each test separately, and by the combination of all of them with respect to the total number of false positives produced by Viola-Jones classifier

Scenarios	Frames	Viola Jones false positives	Test #1	Test #2	Test #3	All
Fixed-camera	11230	122	60,66%	68,85%	53,27%	100%
Moving-camera	27649	563	22,74%	53,46%	93,25%	98,03%

4 Conclusions

In this work we have presented a robust human face detector for TOF cameras that highly reduces the number of false detections, while keeping a very low level of false negative cases. The obtained results demonstrate the interest in using not just intensity images, but also range data to achieve the robustness level demanded by AmI applications. Though nowadays these TOF cameras are still very expensive, we believe that the emergence of these sensors to interface to next generation of interactive games (as Kinect) will put them in the market at a very cheap price. In the future we plan to investigate the use of the proposed method in a service robot aimed to provide AmI capabilities.

References

1. Kolb, A., Barth, E., Koch, R.: ToF Sensors: New Dimensions for Realism and Interactivity. In: CVPR 2008 Workshop on Time-of-Flight-based Computer Vision, TOF-CV (2008)
2. Zhang, C., Zhang, Z.: A Survey of Recent Advances in Face Detection. Technical Report, MSR-TR-2010-66 (June 2010)
3. Hansen, D.W., Larsen, R., Lauze, F.: Improving Face Detection with TOF Cameras. In: Proc. of the IEEE Int. Symposium on Signals, Circuits & Systems (ISSCS), Iasi, Romania, vol. 1, pp. 225–228 (2007)
4. Droeschel, D., May, S., Holz, D., Ploeger, P., Behnke, S.: Robust Ego-Motion Estimation with ToF Cameras. In: Procc of the European Conf. on Mobile Robots (ECMR 2009), Croatia (2009)
5. Gonzalez, J., Galindo, C., Blanco, J.L., Fernandez-Madrigal, J.A., Arevalo, V., Moreno, F.A.: SANCHO, a Fair Host Robot. A Description. In: IEEE Int. Conf. on Mechatronics, ICM 2009 (2009)
6. Böhme, M., Haker, M., Riemer, K., Martinetz, T., Barth, E.: Face Detection Using a Time-of-Flight Camera. In: Kolb, A., Koch, R. (eds.) Dyn3D 2009. LNCS, vol. 5742, pp. 167–176. Springer, Heidelberg (2009)
7. Broxvall, M., Coradeschi, S., Loutfi, A., Saffiotti, A.: An Ecological Approach to Odour Recognition in Intelligent Environments. In: IEEE Int Conf on Robotics and Automation (ICRA 2006), Orlando, Florida (May 2006)
8. Mesa Imagin website, `http://www.mesa-imaging.ch/`
9. Castrillon-Santana, M., Deniz-Suarez, O., Anton-Canalis, L., Lorenzo-Navarro, J.: Face and Facial Feature Detection Evaluation. In: III Int. Conf. on Computer Vision Theory and Applications (2008)
10. Degtyarev, N., Seredin, O.: Comparative testing of face detection algorithms. In: Elmoataz, A., Lezoray, O., Nouboud, F., Mammass, D., Meunier, J. (eds.) ICISP 2010. LNCS, vol. 6134, pp. 200–209. Springer, Heidelberg (2010)
11. Viola, P., Jones, M.: Robust real-time face detection. In: Porc. Int. Conf. on Computer Vision, vol. 57(2), pp. 137–150 (2004)
12. Lienhart, R., Maydt, J.: An extended set of haar-like features for rapid object detection. In: IEEE ICIP 2002, vol. 1, pp. 900–903 (2002)
13. Lange, R., Seitz, P.: Solid-State Time-of-Flight Range Camera. IEEE Journal of Quantum Electronics 37(3) (March 2001)
14. Foix, S., Alenya, G., Andrade-Cetto, J., Torras, C.: Object Modeling using a ToF Camera under an Uncertainty Reduction Approach. In: IEEE Int. Conf on Robotics An Automation (ICRA 2010), Anchorage, Alaska (2010)
15. Freund, Y., Schapire, R.E.: A decision-theoretic generalization of on-line learning and an application to boosting. In: Vitányi, P.M.B. (ed.) EuroCOLT 1995. LNCS, vol. 904, pp. 23–37. Springer, Heidelberg (1995)

Recommendation and Planning through Mobile Devices in Tourism Context

Ricardo Anacleto, Lino Figueiredo, Nuno Luz, Ana Almeida, and Paulo Novais

Abstract. In this paper we present a mobile recommendation and planning system, named PSiS Mobile. It is designed to provide effective support during a tourist visit through context-aware information and recommendations about points of interest, exploiting tourist preferences and context. Designing a tool like this brings several challenges that must be addressed. We discuss how these challenges have been overcame, present the overall system architecture, since this mobile application extends the PSiS project website, and the mobile application architecture.

Keywords: Mobile Recommendation System, Sight Information Provider, Context-Aware, Client-Server Application.

1 Introduction

Mobile systems are becoming popular in the tourism domain, specially due to the pocket size of devices. However, their computational capabilities are still limited compared to a traditional computer. Although in recent years mobile technology has evolved significantly, they still lack performance, especially in battery life, which is the biggest obstacle to mobile performance growth [11]. These limitations must be considered in the creation of any mobile application due to possible technical, ergonomic and economic implications for the user.

There are two types of tourist support applications: mobile information guides (*e.g.*, MultiMundus [6], GeoNotes [4]) and recommendation systems (*e.g.*, Tourism

Ricardo Anacleto · Lino Figueiredo · Nuno Luz · Ana Almeida
GECAD - Knowledge Engineering and Decision Support,
R. Dr. António Bernardino de Almeida, 431. 4200-072 Porto, Portugal
e-mail: {rmao,lbf,nmal,amn}@isep.ipp.pt

Paulo Novais
Universidade do Minho, Campus of Gualtar. 4710-057 Braga, Portugal
e-mail: pjon@di.uminho.pt

P. Novais et al. (Eds.): Ambient Intelligence - Software and Applications, AISC 92, pp. 133–140.
springerlink.com

Information Provider (TIP) [5], m-ToGuide [10], CATIS [8]) [2]. The first ones provide important services to guide the tourist during his travel, displaying points of interest (POIs) according to contextual data. The latter recommend POIs usually according to tourist profiles, allowing the planning and selection of an appropriate trip route based on a set of POIs. Although these systems can be (and should be) integrated, very few approaches integrate both [7]. To improve user-system interaction, mobile devices play an important role since they are able to capture information about the user surroundings (context-awareness) without explicit introduction of these data by the user.

Overall, some tourism support applications like m-ToGuide and Multimundus, have been subject to evaluation with not so positive results. The most complete of the analyzed systems are CATIS and TIP. However, both of them can be improved with some features like booking, augmented reality with 4D representations of sights, and a module to help handicapped people. One aspect that is also very important is the limited capabilities of mobile devices, which none of the described systems take as a main concern.

In order to suppress the limitations found in the analyzed systems, PSiS Mobile was developed. Its foundation is the PSiS (Personalized Sightseeing Tours Recommendation System), which is a tourism decision support system that is capable of recommending points of interest to a tourist according to his personal profile. Besides its recommendation capabilities, it also aids the tourist in planning trips, all through an adaptive web portal [1]. With this implementation of PSiS, tourists can go to the web portal, register and ask for a tour recommendation for a certain region. However, on the field constant assistance for the tourist is not possible since the only mean of interaction with the system is through a web portal. In that sense, a mobile application, namely PSiS Mobile, was necessary to overcome this limitation. Its main purpose is to serve as a personal companion to the tourist. After a plan for a trip is obtained using the web portal features, PSiS Mobile is able to assist the tourist on the field by complementing and updating the plan information according to his current context. Also, context-aware recommendations of sights can be given.

On the next chapter, we present PSiS Mobile and its integration with PSiS, with some detail. Afterwards, we conclude by introducing future work in order to improve PSiS.

2 PSiS Mobile Architecture

PSiS Mobile was developed for the Android platform and it is an occasionally connected application [3], *i.e.*, it does not need to be permanently connected to the Internet in order to work properly. The PSiS Mobile application is divided into four main pieces: ContextService, UserInterface, Communication Manager and Mobile Database.

2.1 *Context Service*

Since Android applications need a service to run a task in background, a Context Service was created. This service manages all the context-awareness modules. These modules are described below:

- **Location Manager**, is responsible for retrieving the user position using the GPS (Global Positioning System);
- **Weather Service**, connects to the WorldWeatherOnline web service to get weather forecasts according to the users position. The system uses it to filter outdoor POIs, *i.e.*, if it is raining, outdoor attractions may not be recommended;
- **Phone Status Manager**, to get mobile device status information. This includes battery status, Internet connectivity (Wi-Fi or 3G) and GPS activity;
- **DateTime Manager**, deals with all date and time information;
- **Planning Service**, incorporates an algorithm to re-plan the original route provided by PSiS. This algorithm was implemented using decision trees;
- **Tracking Service**, tracks user movement and position. For example, it keeps a record of GPS coordinates and for how long the user stays in a certain POI.

2.2 *User Interface*

In Android the user interface is represented by Activities. In PSiS Mobile, each activity represents a single screen in the application, which flow, triggered by user interaction, is depicted in fig. 1.

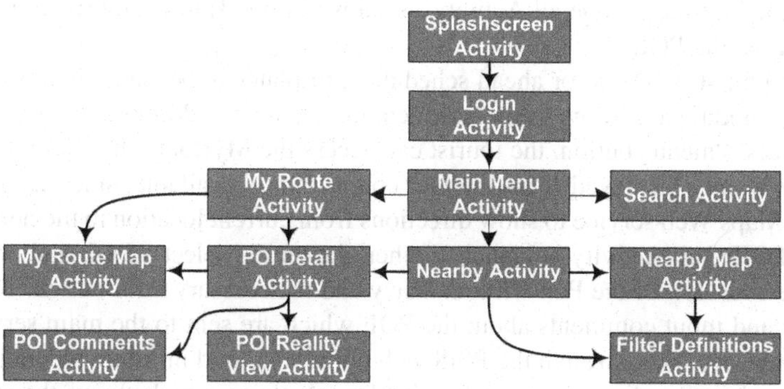

Fig. 1 Application Navigation

The Login Activity (see fig. 2(a)) is the first screen presented that demands user interaction. Here, the user has to log in by introducing an already registered user name and password. To register, he must go to the registration page in the PSiS web portal.

After logging in, the main menu appears. This tab menu is where the user can navigate between the MyRoute Activity, the NearBy Activity and the Search Activity. During the main menu initialization, the context service is launched and begins the context retrieval.

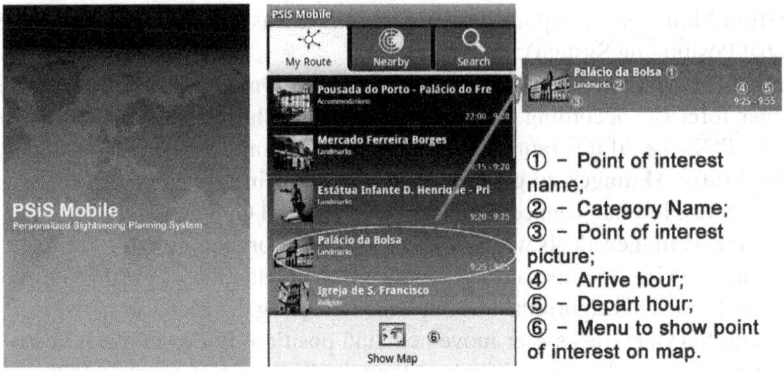

(a) (b)

Fig. 2 a) SplashScreen Activity; b) MyRoute Activity

The MyRoute Activity, presented on fig. 2(b), shows the programmed route, if any, for the current day. It presents the list of POIs to visit along with their arrival and depart hours. As the tourist ends the visit to the POIs, they are excluded from the displayed route and he also can rate them. Optionally, POI detailed information is presented (in the POIDetail Activity, as shown in fig. 3) to the tourist every time he arrives to the POI.

If the tourist is behind or ahead schedule, a re-plan can be suggested by the re-planning module according to the new acquired context. Additionally, by pressing the devices' "menu" button, the tourist can access the MyRouteMap Activity. This Activity is only displayed if an Internet connection is available, since it uses the Google Maps Web service to show directions from current location to the next POI.

The POIDetail Activity is launched when the tourist select a POI, and presents a simple overview of the POI without heavy and unnecessary information. Tourists can rate and input comments about the POI, which are sent to the main server, so they can be accessed through the PSiS web portal and used in future recommendations. Tourists can also access the RealityView Activity, which shows the state of things as they actually exist. Using the mobile device camera, the tourist can see the sight in the device screen and take a picture of it (see fig. 4(b)). Taken photos can be sent to the main server so they are available in the PSiS web portal. Another interesting feature is augmented reality. Using this feature, reality can be augmented by virtual computer layers. In this case, when the user is in front of a POI, more information of that sight is shown, including pictures from other users in different occasions and perspectives. To achieve this, the device must feature a camera, a compass and a GPS receiver.

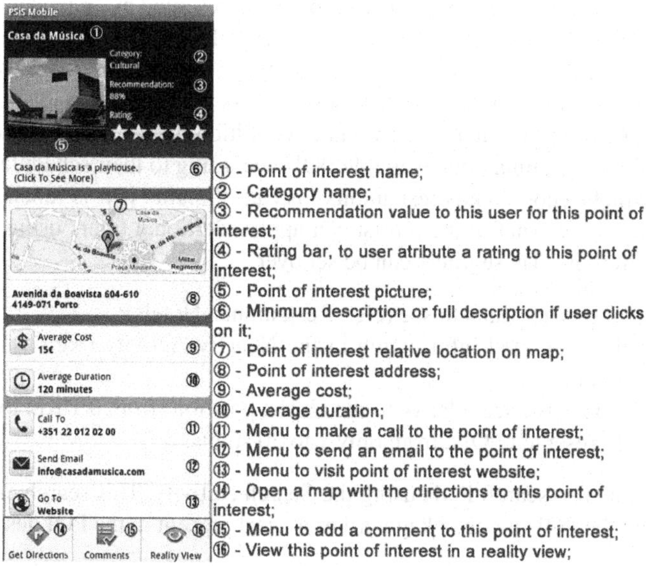

Fig. 3 PoiDetail Activity

(a) (b)

Fig. 4 a) NearbyMap Activity; b) RealityView Activity

The NearBy Activity presents nearby POIs, which are also recommended by the system. The interface is very similar to the MyRoute Activity, showing the recommendation degree and distance to the POI. The distance between tourist location and POIs is calculated using the Haversine formula [9].

The NearByMap Activity, see fig. 4(a), shows NearBy Activity information on a map, where each POI is represented by an icon. The icon corresponds to the POIs' category. For example, the restaurant category is represented by an icon with a knife and a fork. If the tourist clicks on an icon, the POI Detail Activity appears. The current user position is also presented on map through a "person" icon.

PSiS Mobile applies simple content-based filters that rely on context-aware information. The filtering criteria are: weather condition, which filters outdoor POIs, POI schedule, which filters not available POIs according to their schedule, and current hour, which recommends restaurants instead of sights if it is lunch or dinner time. Besides these criteria, the tourist can apply additional filters using the Filter Definitions Activity. These filters can be set over:

- What categories he wants to see (*e.g.*, Restaurants, Hotels);
- Order of the presented information in the NearBy Activity (*e.g.*, by name, by distance, by rating);
- Maximum POI distance relative to the tourist position (from 0.1 Km to 10 Km);
- Amount of presented POIs: from only 1 to 30 sights.

The tourist can also search POIs using the Search Activity. This search can be performed using the POIs name, a keyword or a word present in the POI description.

2.3 Communication Manager and Database

Communication Manager contains the PSiS Mobile DAL, which is a Java library that provides simplified access to data stored in the local SQLite database and in the server. This layer provides an unique abstraction and management over both local and remote data sources. If there are enough local data to show to the user, it only uses the local database. Otherwise, and if an Internet connection is available, a request is made to the server. More information about client-server communications can be seen in section 3.

3 PSiS Middleware Architecture

When the application is installed on the mobile device, an empty database is created. In that sense, the mobile application needs to import information from the server. This is when the middleware plays its role. First of all, when the user logs in and an Internet connection is available, the mobile device requests data to the middleware. Then, the middleware retrieves the necessary data from the server database and sends it back to the mobile device.

The middleware was developed using the Java Servlets API, and is running on a Tomcat server. To exchange information between the mobile application and the middleware, we chose the HTTP (Hypertext Transfer Protocol) protocol along with Googles' Protocol Buffers messages. This module is constituted by six submodules: Database, DAL, Authentication Service, Console, Versions Manager and Communication Manager. The database, although identical to the mobile database, includes three more fields in all tables:

- Date of when the information was sent to the mobile device;
- User identification, to know which user has received the data;
- A Boolean field that represents the necessity or not, to update the data in the mobile device (*e.g.*, if the user generates a new route, this field turns to true in order to indicate that it's necessary to update it in the mobile device).

For all data that is transferred to the mobile device, a copy is saved in the middleware. This way, the heavy task of controlling data versions is left to the Versions Manager sub-module instead of the mobile application.

The third middleware sub-module is a console where all the logs are presented. All requests and a responses are saved in a log table, which can be accessed using a graphical interface (a web page). A DAL is also present to create an abstraction from data present in the middleware and in the server database.

While the Authentication Service ensures data security, the Communication Manager is responsible for handling received/sent messages and redirect them to the appropriate location. Also, if the Internet connection is unstable, the system adapts itself by sending/receiving only one result at a time, instead of a list with many results. This avoids the loss of information.

In fig. 5 a sequence diagram depicts the system steps since the tourist asks for nearby POIs, until he has the result. The tourist starts by going to the NearBy Activity which requests a certain amount of POIs to the DAL. The DAL first inquires the mobile database and, if the desired amount of POIs is not available and an Internet connection is available, it passes the request to the middleware using the Protocol Buffers GetPois message. Then, the middleware sends back the requested amount of POIs using a Protocol Buffers SendPois message. If Internet connection isn't available, the application only shows the POIs present in the mobile database.

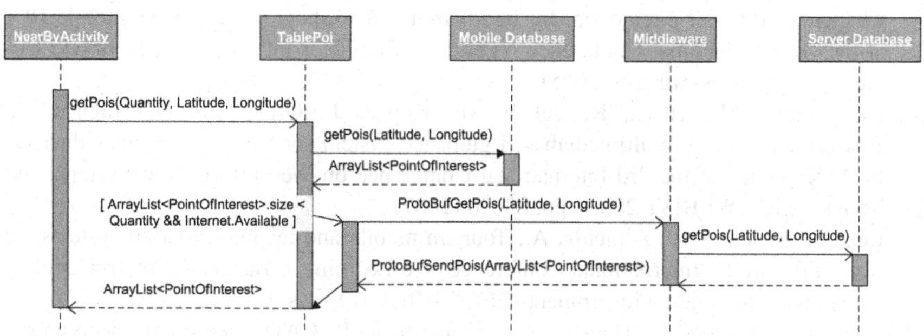

Fig. 5 Sequence Diagram Get Points of Interest

4 Conclusion

Up until now several tourism mobile applications were developed, but they still have several drawbacks and most of them do not account for the multiple limitations of mobile devices. In this paper, we describe the PSiS Mobile architecture, which intends to tackle these limitations. Currently, we are testing it in a real scenario in order to get realistic results to improve the system. Despite mobile device limitations, we want to provide a good user experience, giving tourists a fast and user friendly tool with context-aware adaptation, route planning, augmented reality and built-in social networking features.

In the future, we want to improve the social networking features and create algorithms to interpret the information that is recorded by the mobile application when the tourist is visiting POIs. This will be useful in order to refine even more system recommendations.

Acknowledgements. The authors would like to acknowledge FCT, FEDER, POCTI, POSI, POCI and POSC for their support to GECAD unit, and the project PSIS (PTDC/TRA/72152/2006).

References

1. Almeida, A.: Personalized sightseeing tours recommendation system. In: The 13th World Multi-Conference on Systemics, Cybernetics and Informatics: WMSCI 2009, Florida, USA (2009)
2. Anacleto, R., Luz, N., Figueiredo, L.: Personalized sightseeing tours support using mobile devices. In: World Computer Congress 2010 (WCC 2010), Brisbane, Australia (2010)
3. Anacleto, R., Luz, N., Figueiredo, L.: PSiS mobile. In: International Conference on Wireless Networks (ICWN 2010), Las Vegas, USA (2010)
4. Espinoza, F., Persson, P., Sandin, A., Nyström, H., Cacciatore, E., Bylund, M.: GeoNotes: Social and navigational aspects of location-based information systems. In: Abowd, G.D., Brumitt, B., Shafer, S. (eds.) UbiComp 2001. LNCS, vol. 2201, p. 217. Springer, Heidelberg (2001)
5. Hinze, A., Buchanan, G.: Context-awareness in mobile tourist information systems: Challenges for user interaction. In: International Workshop on Context in mobile HCI at the 7th International Conference on Human Computer Interaction with Mobile Devices and Services, Austria (2005)
6. Kropfberger, M., Tusch, R., Jakab, M., Köpke, J., Ofner, M., Hellwagner, H., Böszörmenyi, L.: A multimediabased guidance system for various consumer devices. In: Proceedings of the 3rd International Conference on Web Information Systems and Technologies (WEBIST 2007), pp. 83–90 (2007)
7. Luz, N., Anacleto, R., Almeida, A.: Tourism mobile and recommendation systems - a state of the art. In: International Conference on e-Learning, e-Business, Enterprise Information Systems, and e-Government (EEE 2010), Las Vegas, USA (2010)
8. Pashtan, A., Blattler, R., Heusser, A., Scheuermann, P.: CATIS: a context-aware tourist information system. In: Proceedings of IMC 2003, 4th International Workshop of Mobile Computing, Rostock, Germany (2003)
9. Robusto, C.: The Cosine-Haversine formula. The American Mathematical Monthly, 38–40 (January 1957)
10. Schneider, J., Schrer, F.: The m-ToGuide Project-Development and deployment of an european mobile tourism guide. In: Evolution of Broadband Services, Eurescom Summer 2003, Heidelberg, Germany (2003)
11. Schuler, R.: Mobile application architecture (2007)

Mobile Personal Health Systems for Patient Self-management: On Pervasive Information Logging and Sharing within Social Networks

Andreas K. Triantafyllidis, Vassilis G. Koutkias, Ioanna Chouvarda, and Nicos Maglaveras

Abstract. Patient self-management is often considered as an important prerequisite towards effective healthcare. This viewpoint has recently been demonstrated by the introduction and adoption of approaches and tools, such as the Personal Health Record (PHR). In the current work, the design of a mobile personal health system for logging information corresponding to the patient status and sharing it within social networks is presented. By utilizing event-driven patterns, the pervasive sharing of the recorded information is enabled, under conditions specified by the mobile user. This "anytime-anywhere" information sharing may be valuable to senders (i.e. patients) and receivers (e.g. relatives, healthcare professionals, similar patients, etc.) in terms of emotional support, mutual understanding, sharing of experiences, seeking of advice and improved self-tracking. A prototype is implemented on a mobile device illustrating the feasibility and applicability of the presented work by adopting unobtrusive health monitoring with a wearable multi-sensing device, a Service Oriented Architecture (SOA) for handling communication issues, and popular micro-blogging services.

Keywords: Personal Health Records, Self-management, Mobile Healthcare Services, Social Networks, Micro-blogging.

1 Introduction

Lately, a number of personal health systems and tools have been demonstrated enabling health information management by the patient himself/herself [1]. Self-management is often regarded as an essential part of efficient disease management, enhancing the patient's role and participation in healthcare services delivery

Andreas K. Triantafyllidis · Vassilis G. Koutkias ·
Ioanna Chouvarda · Nicos Maglaveras
Lab of Medical Informatics, Faculty of Medicine, Aristotle University of Thessaloniki, P.O. Box 323, 54124, Thessaloniki, Greece
e-mail: {atriant,bikout,ioanna,nicmag}@med.auth.gr

P. Novais et al. (Eds.): Ambient Intelligence - Software and Applications, AISC 92, pp. 141–148.
springerlink.com

[2]. Especially, chronic patients may be benefited from self-management activities, in terms of understanding better their disease, enhancing their communication with their doctor, increasing their self-confidence, and so forth [3].

Self-management and quantitative self-tracking have been recently introduced as part of emerging on-line patient communities and social networks like those presented in PatientsLikeME [4]. In such networks, the patient is able to record certain information in regard with his/her health (e.g. a specific health condition) and share it with other patients of the community for purposes of emotional support, exchange of experiences and ideas, education, improved self-tracking etc. Patient willingness to share with others personal health data is a key prerequisite for achieving the afore-mentioned goals. However, the above-mentioned functionality is offered by certain sites requiring constant on-line connectivity, while the integration with health monitoring infrastructures around the mobile user is still in its infancy. In particular, the unobtrusive logging and optional sharing of health information by the mobile users may be of great assistance towards effective (in terms of "anytime–anywhere") and collaborative disease management.

This work presents a novel framework for the construction of mobile personal health systems based on the Personal Health Record (PHR) notion [5], utilizing the acquisition of sensor data from available devices for health monitoring, the recording of health information, and external social networking functionality for sharing personal health information. These systems are particularly targeted at chronic patients throughout their entire everyday activities, who are using portable health monitoring systems, are highly aware of their disease, and may wish to play a more active role in their disease management. The framework supports the configuration of event-driven patterns so as to enable pervasively sharing information within the user's social group. Thus, an environment enabling pervasive and seamless communication between the patient and different actors (e.g. health professionals, relatives, similar patients, etc.) is constructed. A prototype implementation is presented where unobtrusive health monitoring with a wearable multi-sensing device is applied, while a Service Oriented Architecture (SOA) [6] is adopted for the communication among the mobile device, the back-end server and the external social networking platform. Popular micro-blogging services [7] – a form of micro-journalism for posting small pieces of User-Generated Content (UGC) – are utilized in order to demonstrate the social networking functionality.

2 Personal Health Information: Dimensions and Sharing

Personal health information corresponds to the patient status in multiple dimensions as follows:

- ***Vital Sign Measurements & Alerts:*** Health parameters such as the heart rate, respiratory rate, skin temperature and activity are continuously measured by various portable multi-sensing devices [8]. Due to the appeared information overwhelm, event-driven patterns can be initialized by the patient or the health professional, so as to filter the sensed data and record only information of possible value to the patient, as defined according to the configuration of personalized monitoring schemas [9]. For example, an alert of high heart rate may be reported as a result of an average heart rate value within a time-window exceeding a specified threshold.

- ***Health Problems/Symptoms:*** Patients can record various health problems or symptoms met during their daily activities. Examples of such subjective type of information originated from the patient include dizziness, nausea, stress, etc.
- ***Patient Situation:*** Patient situation is manually recorded, e.g. shopping, driving, reading, working, exercising, resting etc., and is characterized by the situation onset and the situation end.
- ***Time and Location:*** Additionally, the time of the day and the location (if available) are crucial parts of contextual information.

Thus, by combining subjective (i.e. patient provided) and objective (i.e. sensor measurements) information, a detailed view of the patient health status is provided for both patients and healthcare professionals. The aggregation of the afore-mentioned information dimensions is considered of particular significance for effective personalized health service delivery. Visualization capabilities of these dimensions during time are offered to the user so as to enable possible discovery of health patterns and improve the self-tracking possibilities in general.

Patients are able to aggregate the diverse recorded building blocks of their personal health information and share it within their social network. Information sharing is taking place either manually or in an event-driven manner. For the latter case, the required conditions for triggering event sharing are encapsulated in the afore-mentioned health information dimensions and are initially configured by the patients according to their needs and preferences. In a next step, they can choose the receivers of the information (e.g. other patients, relatives or health professionals), while finally they can decide on the way of disseminating information choosing between the instant (i.e. information is sent only once) and the continuous mode (i.e. information is always sent, whenever the required conditions are met).

3 System Architecture

In Fig. 1, the overall system architecture is depicted, having the mobile PHR as its core part. The mobile device referred as Mobile Base Unit (MBU) is connected wirelessly with sensors and its *Personal Health Information Controller* regulates various sensor alerts and other types of information dimensions which reflect the patient's status. The MBU is used for recording typical personal health information in the *Personal Health Information Repository*, such as various conditions or problems met, along with various patient situations and alerts (example information recorded by the user is presented in Fig. 2 (a)). All captured information is replicated to the back-end server for safety reasons. Moreover, since the typical mobile device can be still considered as a poor platform for advanced data/information processing, health information persisted in the back-end infrastructure can enable the employment of sophisticated data mining methods for pattern and trend discovery/analysis.

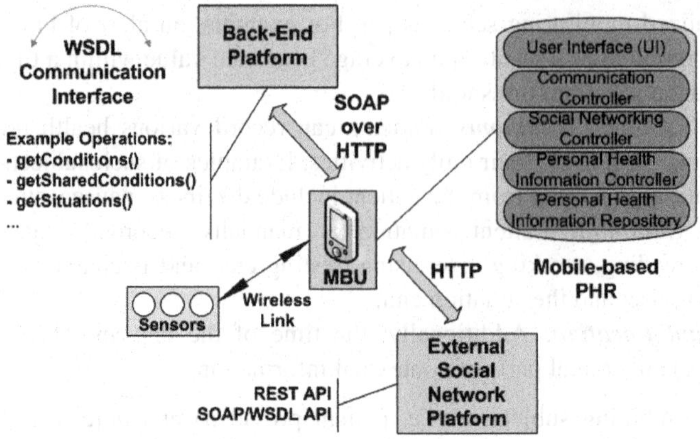

Fig. 1 General overview of the proposed system architecture

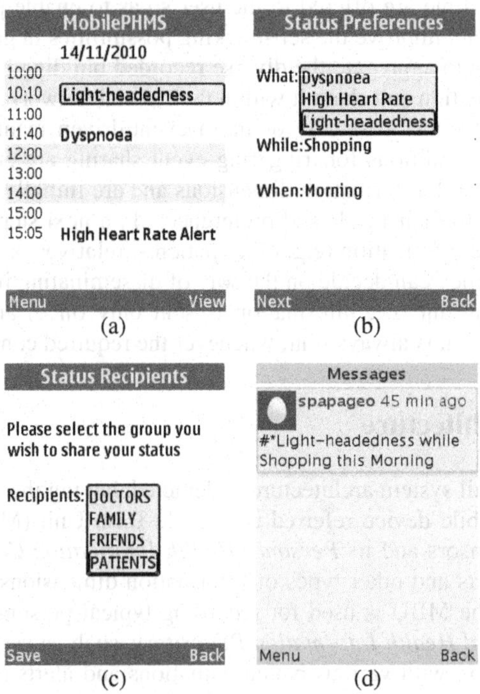

Fig. 2 (a) View of the recorded health conditions in the user's mobile PHR, (b) conditions which trigger status sharing, (c) selected recipients of the shared status, (d) message (sent from user "spapageo") shown to the recipient (via a twitter client application).

Communication between the MBU and the back-end server is achieved via a set of communication interfaces defined and implemented according to the SOA paradigm. SOA has been found to bring significant advantages compared to other architectures, such as interoperability and extensibility [6]. In particular, Simple Object Access Protocol (SOAP)[1] messages over HTTP are transmitted from the MBU, after calling the pre-defined Web service operations related to health information management, e.g. *getConditions()*, *getSituations()*, etc., via communication stubs corresponding to the Web Service Description Language (WSDL)[2] interface. The *Communication Controller* module is responsible for utilizing and controlling the entire client communication with the back-end infrastructure, persisting also unsent information due to potential network unavailability for later transmission.

The MBU communicates with the *External Social Network Platform* via a Representational State Transfer (REST) [10] Application Programming Interface (API) or SOAP/WSDL API, as commonly found in platforms such as Twitter[3]. These APIs provide a way for accessing and using externally the most typical functionalities provided by the platform, while providing also the necessary mechanisms for authentication and privacy via the adoption of protocols such as OAuth[4]. Thus, the MBU may safely connect to the external social network via the design and implementation of appropriate client methods incorporated in the *Social Networking Controller* module.

4 Micro-blogging Services

The proposed system's social networking functionality is realized by utilizing external Web-based micro-blogging services such as those provided by Twitter [11]. In this context, individuals are allowed to construct a public or semi-public profile and articulate social groups within which they share information. UGC is typically in the form of free text messages communicated to subscribers and followers of the message author. Such messages are usually within the limit of 140-160 characters and may optionally provide external links to additional information. In Twitter, a message may additionally be labeled with words followed after a hash so as to ease the message search mechanisms.

In the current implementation, we have elaborated on the event-driven sharing of messages, which are constructed according to the described health information dimensions. In this context, the user is able to create conditional patterns, currently applied in the form of typical IF-THEN rules, according to which the sharing of health information is automated. Initially, the condition (symptoms and alerts) and the situation in which the mobile user is in are manually selected from pre-defined lists (Fig. 2(b)). Alerts originating from sensor data are defined after the initial alert configuration (e.g. thresholds fine-tuning) by the user, as described

[1] http://www.w3.org/TR/soap/
[2] http://www.w3.org/TR/wsdl
[3] http://twitter.com/
[4] http://oauth.net/

in detail in [9]. The resolution of conditions is according to formal medical concepts derived from the Systematized Nomenclature of Medicine - Clinical Terms (SNOMED-CT) terminology with the aid of the API provided by the Unified Medical Language System (UMLS) metathesaurus [12]. Finally, the selection of the particular social group corresponding to the recipients of the shared patient status is taking place, as seen in Fig. 2(c).

Messaging within micro-blogging services is realized by combining predefined tags (information labels) and optional free text (e.g., *#*Light-headedness while Shopping this Morning* as depicted in Fig. 2 (d)). The tags provide a convenient way to discover messages of interest in one's social network. For example, within the constructed messages, the asterisk (*) character is used after the hash, in order to distinguish the condition-related terms provided within our system. Moreover, social analytics and processing may be supported and employed [13], due to the availability of this semi-structured information, without the need for applying complex natural language processing mechanisms. The user is enabled, besides sending UGC with alert, condition or situation-related information, to read messages within a group of subscribers, as sorted by condition or provided after a condition search, making it easy to track messages of interest.

5 Prototype Implementation

A prototype has been implemented on a Nokia N86 smartphone, in order to illustrate the feasibility of the proposed architecture. Java Micro Edition (JavaME) was the chosen development platform, which enabled us to implement and test the described functionality. JavaME provides high-level APIs dealing with the small memory footprint and the limited processing capabilities typically offered by mobile devices. More specifically, in regard with the MBU communication with the back-end, the Java Specification Request (JSR) 172 API was used, in order to provide the Web service functionality based on the SOAP/WSDL approach. In the back-end infrastructure, Apache Tomcat was used as application container and server, MySQL for data persistence and Apache Axis as the underlying Web service engine based on SOAP.

The Zephyr BioHarness[5] physiological monitoring system was used for vital signs monitoring. Zephyr BioHarness is a wearable multi-sensing device incorporating various sensors on a strap placed on the patient's chest for continuous unobtrusive monitoring of the heart rate, activity, posture, respiration rate, and skin temperature. The Zephyr BioHarness provided us with Bluetooth communication capabilities and an API for the transmission of the sensor measurements to the MBU.

The Twitter API[6] relying on REST was used in order to demonstrate the micro-blogging functionality via the MBU. HTTP basic authentication [14] was utilized for authentication purposes, while the kXML[7] package was employed, in order to

[5] http://www.zephyr-technology.com/products/bioharness-bt

[6] http://apiwiki.twitter.com/

[7] http://kobjects.org/kxml/

de-serialize information from the XML-based returned messages of most API calls. For privacy reasons, a private twitter list of people has been created with its subscribers being only the potential system's users.

After the conduction of various performance experiments, Web service invocations from the MBU for communicating with the back-end infrastructure were found to last on average 1.45 sec until reception of the response. The Twitter API calls lasted on average 1.9 sec for message transmission to the social group while the reading of the subscribers' messages within the list lasted on average 4.2 sec for 10 new messages with XML de-serialization time included in the final result. Thus, system performance was found to be sufficient enough in terms of communication and XML processing cost, although further tests are needed to fully explore all the performance evaluation aspects.

6 Conclusion

This paper proposed an approach towards chronic patients' self-management based on a mobile personal system encapsulating services to support patients in health information management and sharing. The primary focus of this work was on the implementation of a mobile solution to achieve pervasive and seamless communication among patients and their networked community.

At the current stage, our prototype implementation constituted a technical proof-of-concept as regards the feasibility and applicability of the proposed approach. It is evident that the presented system is primarily targeting at patients willing to play a more active role in managing their disease. The ultimate goal of this approach is further enhancing the patient's personal role in healthcare information management and promoting collaborative healthcare.

Privacy policies in regard with protecting personal health data need to be further explored [15] while evaluation of the presented system is necessary to assess user acceptance, as well as the extent of its contribution in patient self-management. Moreover, our future work involves the further development of methodologies for handling contextual data, behavioral monitoring based on user-to-system interactions, and appropriate methods for the collaborative filtering of information and discovery of patterns.

Acknowledgments. The research leading to these results has received funding from the Ambient Assisted Living (AAL) Joint Programme under Grant Agreement n° AAL-2008-1-147 – the REMOTE project (http://www.remote-project.eu/).

References

1. Mattila, E., et al.: Empowering citizens for well-being and chronic disease management with wellness diary. IEEE Trans. Inf. Technol. Biomed 14(2), 456–463 (2010)
2. Mosen, D.M., Schmittdiel, J., Hibbard, J., Sobel, D., Remmers, C., Bellows, J.: Is patient activation associated with outcomes of care for adults with chronic conditions? J. Ambul Care Manage. 30(1), 21–29 (2007)

3. Lorig, K.R., Sobel, D.S., Ritter, P.L., Laurent, D., Hobbs, M.: Effect of a self-management program on patients with chronic disease. Eff. Clin. Pract. 4(6), 256–262 (2001)
4. Frost, J.H., Massagli, M.P.: Social uses of personal health information within PatientsLikeMe, an online patient community: What can happen when patients have access to one another's data. J. Med. Internet Res. 10:e15 (2008)
5. Tang, P.C., Ash, J.S., Bates, D.W., Overhage, J.M., Sands, D.Z.: Personal health records: Definitions, benefits, and strategies for over-coming barriers to adoption. J. Am. Med. Inform. Assoc. 13(2), 121–126 (2006)
6. Singh, M.P., Huhns, M.N.: Service-oriented computing: Semantics, processes, agents. J. Wiley and Sons, Chichester (2005)
7. Ebner, M., Schiefner, M.: Microblogging - more than fun? In: Proceedings of IADIS Mobile Learning Conference, pp. 155–159 (2008)
8. Konstantas, D.: An overview of wearable and implantable medical sensors. IMIA Yearbook 2007 2(1), 66–69 (2007)
9. Triantafyllidis, A., Koutkias, V., Chouvarda, I., Maglaveras, N.: An open and reconfigurable wireless sensor network for pervasive health monitoring. Methods Inf. Med. 47(3), 229–234 (2008)
10. Leonard Richardson, S.R.: RESTFul web services. O'Reilly, Sebastopol (2007)
11. Java, A., Song, X., Finin, T., Tseng, B.: Why we twitter: Understanding microblogging usage and communities. In: Proceedings of the Joint 9th WEBKDD and 1st SNA-KDD Workshop (2007)
12. Bangalore, A., Thorn, K.E., Tilley, C., Peters, L.: The UMLS Knowledge Source Server: An object model for delivering UMLS data. In: Proceedings of AMIA Annual Symposium, pp. 51–55 (2003)
13. Kleinberg, J.M.: Challenges in mining social network data: processes, privacy, and paradoxes. In: Proceedings of the 13th ACM SIGKDD, pp. 4–5 (2007)
14. Franks, J., Hallam-Baker, P., Hostetler, J., Lawrence, S., Leach, P., Luoto-nene, A., Stewart, L.: HTTP authentication: Basic and digest access authentication. IETF RFC2617 (1999)
15. Martino, L., Ahuja, S.: Privacy policies of personal health records: An evaluation of their effectiveness in protecting patient information. In: Proceedings of the 1st ACM International Health Informatics Symposium, pp. 191–200 (2010)

Multi-Agent Strategy Synthesis in Smart Meeting Environments

René Leistikow

Abstract. A significant challenge of Ambient Intelligence (AmI) systems and applications, such as smart meeting environments, is how to assist the user unobtrusively by using ubiquitous information technologies and computing capabilities. Furthermore, the user should be able to integrate his personal mobile devices into the existing device ensemble of the environment to create a coherent ad-hoc ensemble. Hence, these systems require inter alia a dynamic strategy synthesis that fulfills inferred user's intention and integrates components seamlessly. Multi-Agent systems support such kind of dynamic system behavior, so that we assume that it is a suitable paradigm to model strategy synthesis in these environments.

In spite of several solutions for different planning problems in smart meeting environments, a concrete domain specification and a combination of these solutions are still unavailable. This doctoral paper describes the current research and explains how the author's PhD topic fits into this research area.

Keywords: strategy synthesis, planning domain, smart meeting environments, multi-agent system.

1 Introduction

> "The real power of the concept comes not from any one of these devices - it emerges from the interaction of all of them." [19]

Problem. There exists numerous proposals for addressing planning and device cooperation in smart meeting environments, but, to the best of our knowledge, a concrete planning domain specification is still unavailable. Today, there is no unified, comprehensive catalogue defining the set of planning problems that have to be considered in smart environments. Therefore, there is only little knowledge on how

René Leistikow
MMIS, Computer Science, University of Rostock, Germany
e-mail: `rene.leistikow@uni-rostock.de`

P. Novais et al. (Eds.): Ambient Intelligence - Software and Applications, AISC 92, pp. 149–155.
springerlink.com

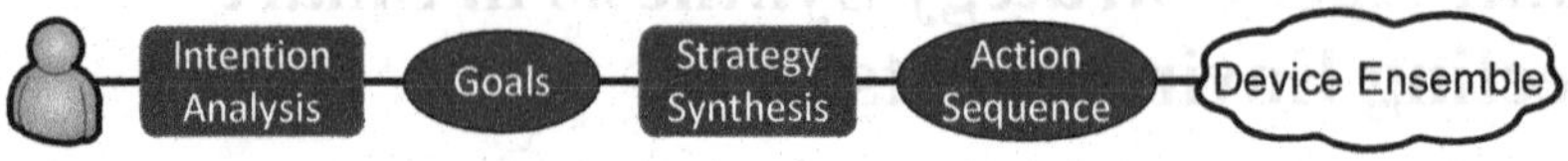

Fig. 1 Goal-based user assistance

different subproblems in smart environment control interact and how they can be captured within a unified modeling approach. However, without a unified model it is not possible to identify, let alone to resolve, interactions and conflicts between different subproblems. In addition, specific algorithmic strategies have proven to be effective in solving specific device cooperation problems in smart environments. However, it is not very well understood, how to combine these different strategies (e.g., causal planning on the one hand, resource optimization on the other), within a unified algorithmic framework.

2 Background

Goal-based User Interaction. In our group, we pursue a goal-based interaction approach (cp. Figure 1) instead of a function-oriented approach to realize Mark Weiser's vision of an unobtrusive user assistance in a ubiquitous computing environment [19]. The reason for this choice is that when people are using a technical infrastructure they have goals they want to achieve. Thus, they do not want to think about concrete device functions [10].

In the goal-based interaction model, the intention analysis initially tries to infer the user's goals from perceived (sensor) data (e.g., via speech recognition, user's position) using, typically, artificial intelligence approaches like dynamic Bayesian networks [9]. Accordingly, the inferred user's goals have to be transformed / decomposed into (achievable) goals of the environment. This process is called *goal refinement*[1]. Then, the strategy synthesis performs means-end reasoning by using goals of the environment as well as domain (e.g., device descriptions) and perceived world knowledge (e.g., temperature or luminosity). The data sets are, typically, encoded in PDDL [8], using STRIPS operators [6] with preconditions and effects for each device function. The solution of this planning process is an action sequence, which can be executed in the last step by instructing devices of the ensemble. In this paper, we focus on the strategy synthesis and action sequence generation in smart meeting environments.

SmartLab. A smart meeting environment, called Smart Appliance Lab (SmartLab), serves as an experimental infrastructure for our research (cp. Figure 2). This room is equipped with sensing devices, such as luminosity sensors, real-time location systems (active badge system and sensitive floor tiles), motion sensors and cameras; and actuators, such as (steerable) projectors, motor blinds and motor screens. In addition, devices are connected and controlled via the European Installation Bus

[1] Please notice that this process is not explicitly shown in Figure 1.

Fig. 2 Unversity of Rostock's Smart Appliance Lab (SmartLab)

(EIB) or KNX respectively[2]. To simplify the access, devices and their capabilities are encapsulated in Java objects. Moreover, these objects can be published either via network or via a specialized tuple space middleware [1].

3 Planning Domain

In this section, we identify a selection of the numerous different planning issues in smart meeting environments to gain a more detailed understanding of this domain.

Temperature and Luminosity Control. First, it might be desirable for the smart meeting environment to provide a pleasant room atmosphere (e.g., adjust lighting conditions, regulate heating system), which needs to be adjusted to the individual requirements of users. [3, 10]

Recording Meetings. For documentation, archiving and indexing of meeting results, it could be helpful to record the meeting using video cameras and microphones. [2, 14]

Distributed Meeting. A distributed meeting involves people in smart meeting room as well as other places, possibly in different countries, with access to different devices. [13]

---

[2] `http://www.knx.org/`

Display Mapping Problem (DMP). "So called Multi-Display Environments [e.g., SmartLab] support collaborative problem solving and teamwork by providing multiple display surfaces for presenting information. [...] One difficult task here is the Display Mapping problem - that is, deciding which information to present on what display in order to optimally satisfy the users' needs for information." [10]

Unfortunately, the DMP belongs to the class of Quadratic Assignment Problems, which have a NP-hard computational complexity.

4 Planning Algorithms

In this section we present two exemplary strategy synthesis approaches which were developed by our colleagues and are aimed to solve the DMP as one example, among others, of a sophisticated planning problem in a smart meeting environment. Therefore, we classify these techniques into two paradigms of distributed planning, introduced by Durfee in [5].

Centralized Planning for Distributed Plans. The first paradigm describes a planning system which develops a plan for a group of agents in a centralized manner. The central instance gathers world and domain knowledge, plans and distributes parts of the plan to the agents. Additionally, it is responsible for a synchronized execution of the device actions.

In [12], Marquardt et. al. pursued a centralized classical AI planning approach. They implemented a central component called *Composer*, which consists of three main functional units. For gathering information (e.g., user's intentions, world state, device descriptions) a *domain assembler* is used. In addition, it is able to create a planning problem by using the gathered data sets. The *plan selection and monitoring* unit wraps and controls different general-purpose AI planners (e.g., LPG, UCPOP or SGP). This unit sends the planning problem to the controlled planners, which try to solve a planning problem in a simultaneous manner. The *Composer* selects the solution of the fastest planner, validates the plan using the *plan validation* unit and distributes the parts of the plan to device-controlling agents.

Distributed Planning for Distributed Plans. The second paradigm describes how a group of agents can cooperate with each other to form individual plans, while dynamically coordinating their activities along the way. In this case the agents may be *selfish* which means that each agent wants to maximize its own utility value instead of maximizing the social welfare utility. Thus, it is possible that coordination problems arise (e.g., resource conflicts) and agents may need to negotiate to solve these conflict situations. It should be noted that this is the most sophisticated, but also the most flexible and robust, paradigm.

In [16, 17], Plociennik et. al. suggested a decentralized extension of Pattie Maes' action selection algorithm [11] for the use in smart ad-hoc environments. In the original algorithm an autonomous agent consists of a set of *competence modules*. The relationship between the *competence modules*, or, to be more precise, the relationship between the preconditions and effects of the *competence modules* can be expressed by several types of links (e.g., successor, predecessor and conflicter links). Linked *competence modules* are able to activate and / or to inhibit each other by exchanging energy via different links. Therefore, the action selection is an emergent property of this network of *competence modules* which means that the *competence module* with the highest amount of energy will be executed.

Nonetheless, it should be noted that there are other methods which deal with planning problems in smart environments. In the majority of cases, these approaches are based either on condition-based rule systems (e.g., Microsoft's EasyLiving Project [3]) or plan recognition methods (e.g., Intelligent Classroom Project [7]). A further detailed description would go beyond the scope of this paper.

5 Main Issues

Below, the main open issues which have to be resolved according to reach the overall goal of an multi-agent unified algorithmic framework for the strategy synthesis in smart meeting environments are mentioned[3].

1. **Definition of a Domain Specification.** As mentioned above, a concrete planning domain specification for smart meeting environments is still unavailable. However, such a description is essential, for example, for the goal refinement process (see section 1) and for a better understanding of the application domain "smart meeting environment". In section 3, we identify four different planning issues in smart meeting environments.
2. **Identification of Planning Subproblems.** With a concrete domain specification, we assume that it is possible to identify independent subproblems of the planning domain, based on the assumption that it is easier to solve many small problems than one big problem. The temperature, for example, could be controlled independently from lighting conditions.
3. **Selection of Planning Algorithms.** The division of the planning domain allows to choose suitable algorithms and heuristics for each individual subproblem. That requires an extensive evaluation of suitable algorithms for each subproblem. For instance, we expect that the usage of (lightweight) reactive behaving agents instead of a (heavyweight) deliberative planning process should be sufficient enough to solve several problems (e.g., adjusting lighting conditions). For instance, the DMP could be solved with one of the algorithms mentioned above (see section 4).
4. **Finding Optimal Solutions for the Planning Problems.** Moreover, it is possibly necessary to define metrics for planning problems to decide which plan is the most appropriate one in a given situation or, to be more precise, leads to the highest utility value for the ensemble in terms of optimal plan generation. It should be noticed that users also accept suboptimal solutions in some situations [15].

[3] These issues are preliminary and could be changed in the running research process.

5. **Dealing with Conflicting User Goals.** In [4], Cook et. al. identified the problem of conflicting user goals. To deal with this problem, we want to create a hierarchical model which is able to inhibit action executions by prioritization of subplans and their related actions to resolve the conflicts.

In addition to the main issues mentioned above, other problems arise in the strategy synthesis' deployment process itself. For example, since we want a distributed planning and execution process (based on a multi-agent system), we have to take the communication efforts for negotiation into account. More detailed descriptions of these issues can be found in [18].

6 Conclusion and Future Work

In this doctoral paper we described a project that aims to optimize existing strategy synthesis methods and algorithms in dynamic ad-hoc environments particularly smart meeting environments. Therefore, the primary goal is the creation of a multi-agent unified architecture for strategy synthesis in smart meeting environments, which is able to identify and to solve planning problems using user intentions, domain and world knowledge. In this process, planning problems will be decomposed into subproblems. Then, the latters will be handed over to specialized and optimized planning algorithms, which generate action sequences for the ensemble.

With respect to the early state of the research, only the idea for further investigation on this topic is defined.

Based on predefined possible intentions of users, we now start to decompose these intentions into achievable goals of the environment to gain a precise understanding of the described application domain, to give a formal definition of the planning domain and to derive subproblems.

Acknowledgements. This work was funded by the German Research Foundation (DFG) within the Graduate School 1424 (MuSAMA).

References

1. Bader, S., Ruscher, G., Kirste, T.: Decoupling smart environments. In: Bader, S., Kirste, T., Griswold, W.G., Martens, A. (eds.) Proceedings of PerEd2010, Copenhagen (September 2010)
2. Brotherton, J.A., Abowd, G.D.: Lessons learned from eclass: Assessing automated capture and access in the classroom. ACM Trans. Comput.-Hum. Interact. 11, 121–155 (2004)
3. Brumitt, B., Meyers, B., Krumm, J., Kern, A., Shafer, S.: EasyLiving: Technologies for intelligent environments. In: Thomas, P., Gellersen, H.-W. (eds.) HUC 2000. LNCS, vol. 1927, pp. 97–119. Springer, Heidelberg (2000)
4. Cook, D.J., Das, S.K.: S.: How smart are our environments? an updated look at the state of the art. Pervasive and Mobile Computing 3, 53–73 (2007)

5. Durfee, E.H.: Distributed problem solving and planning, pp. 121–164. MIT Press, Cambridge (1999)
6. Fikes, R.E., Nilsson, N.J.: Strips: A new approach to the application of theorem proving to problem solving. Artificial Intelligence 2(3-4), 189–208 (1971)
7. Franklin, D.: Cooperating with people: the intelligent classroom. In: Proceedings of Fifteenth National Conference on Artificial Intelligence, pp. 555–560 (1998)
8. Ghallab, M., Nationale, E., Aeronautiques, C., Isi, C.K., Penberthy, S., Smith, D.E., Sun, Y., Weld, D.: Pddl - the planning domain definition language. Technical report (1998)
9. Giersich, M., Kirste, T.: Effects of agendas on model-based intention inference of cooperative teams. In: International Conference on Collaborative Computing: Networking, Applications and Worksharing, CollaborateCom 2007, pp. 456–463 (November 2007)
10. Heider, T.: A Unified Distributed System Architecture for Goal-based Interaction with Smart Environments. PhD thesis, University of Rostock (2009)
11. Maes, P.: How to do the right thing. Connection Science Journal 1, 291–323 (1989)
12. Marquardt, F., Uhrmacher, A.M.: An ai-planning based service composition architecture for ambient intelligence. In: Workshop Proceedings of the 5th International Conference on Intelligent Environments, pp. 145–152 (July 2009)
13. Marreiros, G., Santos, R., Ramos, C., Neves, J., Novais, P., Machado, J., Bulas-Cruz, J.: Ambient intelligence in emotion based ubiquitous decision making. In: Artificial Intelligence Techniques for Ambient Intelligence, Hyderabad, India (January 2007)
14. Moore, D.: The idiap smart meeting room. Idiap-Com Idiap-Com-07-2002, IDIAP (2002)
15. Plociennik, C., Wandke, H., Kirste, T.: What influences user acceptance of ad-hoc assistance systems? - a quantitative study. In: MMS, pp. 57–70 (2010)
16. Reisse, C., Kirste, T.: A distributed action selection mechanism for device cooperation in smart environments. In: IET 4th International Conference on Intelligent Environments, pp. 1–8 (July 2008)
17. Reisse, C., Kirste, T.: A distributed mechanism for device cooperation in smart environments (2008)
18. Satyanarayanan, M.: Pervasive computing: Vision and challenges. IEEE Personal Communications 8, 10–17 (2001)
19. Weiser, M.: The computer for the 21st century. SIGMOBILE Mob. Comput. Commun. Rev. 3, 3–11 (1999)

Reusable Gestures for Interacting with Ambient Displays in Unfamiliar Environments

Radu-Daniel Vatavu

Abstract. Gesture-based interfaces offer great opportunity for achieving natural and intuitive interactions in the age of ambient intelligence. Therefore, many efforts have been dedicated in the recent years for proposing gesture acquisition, recognition, and interaction techniques. The age of ambient intelligence has also incorporated gestures into practical applications having as main goals adaptive and personalized interactions. However, practitioners are faced with several issues when implementing gesture-based interfaces for public displays. Let alone the acquisition technologies and recognition algorithms, there are currently little to no rules for creating the specific set of gestures for a given application. Therefore, a compromise is usually made by associating gestures and functions based on the designer's expertize and experience. This contradicts the goals of ambient intelligence where the interaction should be personalized to each user. With this respect, we propose a novel concept for reusing a set of user-defined and user-specific gestures for interacting with public ambient displays. The concept relies on an important shift of perspective: it is not the users that adapt to the interface but instead the interface employs the users' own gesture set with their own preferred meanings.

Keywords: public displays, interaction, ambient displays, ambient interfaces, gestures, sensor glove, hand postures, gesture recognition.

1 Introduction

Gesture-based interfaces have found their way into consumer applications being now common for mobile devices [21], computer games [5], interactive surfaces [6], and large displays [2, 3, 4]. The main motivation for using gestures in the interface lies with them being natural and familiar for users to perform. Gestures also

Radu-Daniel Vatavu
University Stefan cel Mare of Suceava, 13, Universitatii, Suceava 720229, Romania
e-mail: vatavu@eed.usv.ro
http://www.eed.usv.ro/~vatavu

P. Novais et al. (Eds.): Ambient Intelligence - Software and Applications, AISC 92, pp. 157–164.
springerlink.com

act as excellent shortcut commands leading to efficient interactions and small task completion times for users having attained expert performance [1].

Implementing the technology for detecting and understanding gestures represents a considerable challenge for ambient intelligence researchers interested in practical applications. Although gestures bring many advantages for the interaction process, the designers of such interfaces are faced with considerable problems. Let alone the algorithms complexity as well as the processing power needed for real-time non-invasive recognition of free-hand or whole-body gestures [9, 14], special care needs to be devoted to designing the gesture set. The main problems are represented by finding gestures that are easy to execute, learn, and recall [11].

As more and more interfaces will presumably include gesture commands, especially in the age of ambient intelligence and ambient displays, a variety of gestures are likely to be proposed for the same function by each application designer. Users will find difficult to recall gesture commands which will certainly create confusion while they move from one ambient display to another. Even more, different users will have different gesture preferences for specific actions. This has already been observed by researchers eliciting gestures from non-technical users [20]. The problem is especially important in the context and vision of gesture-controlled ambient displays for which the interaction is usually brief, to the point, and time- and location-opportunistic.

This paper introduces the concept of user-specific gestures that can be reused for new unseen ambient displays in unknown environments. The idea is to have users carry their preferred and practiced gesture sets on some storage device (the mobile phone being a good example) and upload them to the public display they are about to use. The interface will then associate the users' own gestures with the tasks available in the system by matching them with users' preferred function associations. The advantages are two-fold: firstly, the robustness and accuracy of the recognition algorithm the system employs will considerably improve (as the system employs the users' own executions as training examples acting as if user-specific training had just occurred); secondly, users already know how to interact with the system by reusing their own gestures for performing actions for which they have already performed solid mental associations. The goal is to promote user-specific gesture sets to be reused for interacting with unknown ambient interfaces, building thus one of the goals of ambient intelligence: user personalization.

2 Related Work

We focus on related works that address the problem of interacting with public ambient displays in order to highlight the common design practices as well as to identify the problems designers are currently facing. Implementing a successful gesture-based interface requires selecting a technology for acquiring gestures, using a recognizer, and designing a set of gesture commands. Each of these aspects brings in considerable challenges that relate to the robustness of acquisition/recognition and efficiency/effectiveness of the interaction [17].

The most common technologies that have been proposed for interacting with public ambient displays include mobile phones [2, 3, 4] and gesture recognition [7, 8, 10, 16, 19]. Mobile phones have been primarily used for controlling cursors on remote displays in a WIMP-like fashion [2, 3, 4] but also as sensing devices for capturing gestures such as tilting and throwing [4]. Beyond mobile devices, various technologies have been used in order to capture gestures for interacting with ambient displays. Vogel and Balakrishnan [19] explored the use of free-hand interaction by employing a sensor data glove and proposed several pointing and clicking techniques. Malik et al. [8] used computer vision in order to capture and recognize postures of the users' hands above a tabletop. Shoemaker et al. [16] introduced the shadow reaching technique which uses a perspective projection of the user's shadow on the remote display for the purpose of easy access over the large area of the display.

Although many studies have been devoting to developing new acquisition, recognition, or interaction techniques, few have explored the design of the gesture set. The current practice lets designers propose gesture commands based on their previous expertise and experience. However, recent studies have shown that users will enter different gestures for the same command/task. For example, Wobbrock et al. [20] elicited tabletop gestures from non-technical users by portraying the effect of a command and asking the gesture that could trigger the portrayed effect. They found that users rarely care about the number of fingers they use; that one and two hands gestures can be used for the same command; and that there are commands for which little agreement can be observed between users. This suggests the need for personalizing the user interface in order to accommodate one's specific gestures as different users prefer different gesture commands.

3 Reusable Gestures

The discussion from the previous sections has identified two common problems encountered when implementing gesture-based interfaces:

- the designer's match between gestures and actions is not always intuitive to everyone and different users will prefer different gestures for the same action. This has been clearly showed by Wobbrock et al. [20] in their experiments;
- user-dependent is preferred to user-independent training as gesture recognizers will perform more accurately. However, going through a training stage each time when interacting with a new system poses problems with respect to the fluidity of the interaction. This is specifically important when on-the-run consumers need to interact with multiple displays while moving in a world where ambient displays are likely to become prevalent.

A new solution is proposed that addresses both inconveniences by having users employ their preferred gesture set each time when interacting with a public ambient display. The preferred gestures are being carried on the mobile device and uploaded to the interactive system before interaction takes place. This solves gracefully both the problems mentioned above, assuring a perfect match between gesture and function for each user as well as robust recognizer performance due to user-specific training (as users are uploading their own training samples). The gesture set can be stored on the mobile phone in the form of a data file containing the association between gestures and functionality as well as training samples for each gesture type. Users will connect to the public display via the available wireless connection of the device (Bluetooth, wireless LAN, etc.) and simply transfer the data file to the ambient display. Figure 1 illustrates the concept.

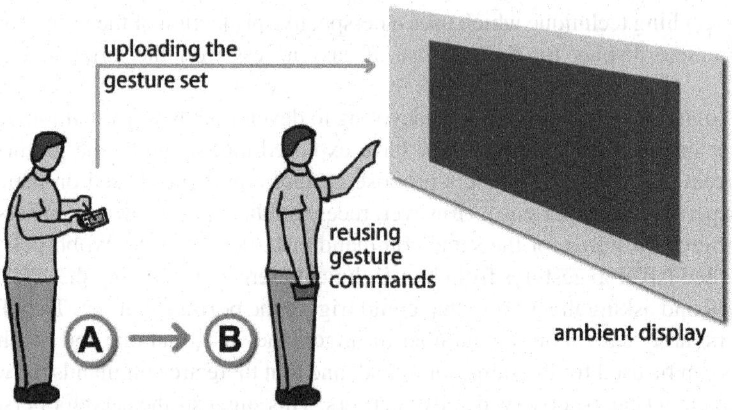

Fig. 1 Interacting with reusable gestures: (A) the preferred/practiced gesture set is transferred to the interface of the public ambient display before interaction takes place (B)

3.1 Phone vs. Free-Hand Gestures

As discussed in the related work section, previous solutions for interacting with remote displays either use mobile phones or gestures. The proposed concept is actually a mix of the two: the mobile phone represents the storage medium for user-specific data while actual interaction is achieved using gestures. There are no constraints however on how these gestures should be captured and they may be very well acquired using the mobile device acting as a sensor [3, 4]. It is interesting therefore to analyze the differences between the two major approaches being previously considered for controlling content on public displays: mobile phones vs. gestures. In some cases, the two approaches overlap such is the case of using tilt or throw movements [3, 4] captured by the sensing capabilities of the mobile phone. Each approach presents advantages but also weaknesses for both designers and practitioners that built such systems but also for users that employ them. Otherwise put, the technology is already here as it has been proven, be it gesture-based or phone-enabled but the question still remains of how to use it in order to achieve intuitive and fluent interactions. Table 1 lists a few important problems which are present at different levels for both approaches.

Table 1 Phone vs. gesture-based interaction with public ambient displays

	Phones	Gestures
1.	Require establishing a connection (wireless LAN, Bluetooth, IR) which may take time and therefore affect the fluidity of the interaction (this also influences serendipity which means the users' ability to spontaneously interact with a large display)	Require reliable tracking technology for detecting users and their actions. False positives demand re-entering of commands which again affects the fluidity of the interaction
2.	May need downloading and installation of additional software	Unless told and practiced before, how can users know and master the gestures that the public ambient display exposes?
3.	Privacy concerns (related to phone data and person location)	Privacy concerns (as the face and user's actions may be captured by video cameras)
4.	Social acceptance: willingness of interacting in public	Social acceptance: willingness of performing gestures in public

The introduced concept touches all the points discussed in Table 1 as it uses both the mobile phone as well as gestures. Solutions for the serendipity problem (the 1*st* point in the table) can be addressed by making the phone automatically connect to near-by networks and investigate available services. False positives in detecting gestures can be tighter controlled as computer vision becomes more and more mature as a field. The concept addresses nicely the 2*nd* problem by transferring user-specific gestures to the system: users already know what the gestures are. The privacy concerns regarding the mobile phone are also likely to be overcome as network security technology will provide trustworthy communications. Interestingly, gestures have been proposed as authentication schemes for public displays [12]. The social acceptance represents however an important issue to overcome. Not only that mobile phones are expressly banned in some areas or in some contexts (trains, hospitals, etc.) but people tend to hesitate using interactive gesture commands in front of a public audience [15].

3.2 Discussion of a Practical Implementation

In order to demonstrate and further discuss the concept, a gesture-based interface is considered that allows users to access the information provided by an ambient display. The functions that the display exposes include: select a graphical item; expand the item (resulting in more information being displayed; open or maximize represent equivalent options); close the expanded item (thus reverting back to the initial arrangement of items; minimize and main menu represent equivalent names for this option).

Gestures are captured using the 5DT Data 5 Ultra glove[1] equipped with 5 optical sensors measuring the flexion of each finger. The glove conveniently allows for rapid prototyping of the gesture-based interface in order to demonstrate the concept. As previous work also considers [19], as computer vision becomes mature as a field, free-hand gesture acquisition will become robust removing the need for worn equipment. The glove assures therefore rapid implementation with robust results. A hand posture p is represented using five flexion values $p = (f_1, f_2, f_3, f_4, f_5)$, one for each finger. A straightforward way to recognize postures is to use the nearest-neighbor classifier working with the Euclidean metric: given a set of n examples $p_i = (f_1^i, f_2^i, f_3^i, f_4^i, f_5^i)$, $i = 1..n$, for which the class information is already available ($p_i \in C_j$, $j = 1..m$), a new hand posture p is recognized as belonging to the class C_k of its closest sample in the set:

$$p \in C_k \Leftarrow k = \arg\min_{i=1..n} \|p - p_i\| \tag{1}$$

where $\|p - p_i\| = \left(\sum_{j=1}^{5} (f_j - f_j^i)^2 \right)^{\frac{1}{2}}$ represents the Euclidean distance between hand postures p and p_i. We omit more details of the recognition process cause of space constraints and instead refer the reader to similar approaches [13, 18].

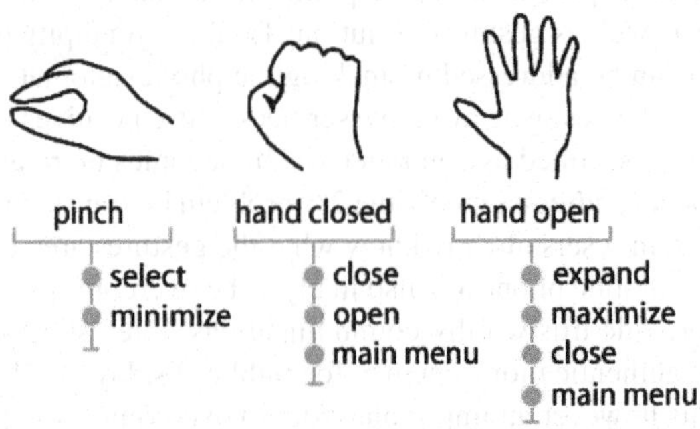

Fig. 2 The set of hand postures designed for the glove-based prototype. The postures are simple enough to recall and perform yet they may have different function associations for different users.

What is interesting instead is that different users can have different preferences when assigning meaning to gestures. Figure 2 illustrates the proposed gesture set (3 classes C_j) for the functions of the considered prototype where each hand posture may have different (and perfectly intuitive and thus valid) function associations. For example, the *pinch* posture can be used to minimize but also to select, while *hand open* can trigger the expand or close functions. At the same time, returning back to the main menu can be very well performed by both *hand closed* and *hand*

[1] http://www.5dt.com/products/pdataglove5u.html

open postures. Data captured from each user can be simply stored as an XML file containing all the p_i training samples together with their class information and associated meaning (such as open, close, minimize, etc.) This file will be transferred to the ambient display using the available connections of the mobile device.

It is important to note that the gesture data file must already exist which implies that a training procedure already took place. This can happen at the user's pace (in their own home environment for example) and it can be more than a one-time procedure: it should continue each time new gestures are being added for new functions or when existing gestures are being removed or updated. We consider this to be the user-specific training that takes place at the user's own rhythm independently from the actual interaction in the ambient environment.

4 Conclusions

The paper introduced a new concept for reusing gestures in unfamiliar environments. The concept can be easily implemented as a reusable set of gestures that are carried by users on their mobile phones and which are being uploaded to the ambient display prior to interaction. Future work will focus on performing user studies in order to understand the applicability of the concept, its acceptance, and the perceived experience. The serendipity challenge should also be properly addressed. The solution opens up the way to personalized interaction with unfamiliar interfaces and unknown environments by building on the concept of adaptation and personalization of ambient intelligence.

Acknowledgements. This paper was supported by the project "Progress and development through post-doctoral research and innovation in engineering and applied sciences- PRiDE - Contract no. POSDRU/89/1.5/S/57083", project co-funded from European Social Fund through Sectorial Operational Program Human Resources 2007-2013.

References

1. Appert, C., Zhai, S.: Using strokes as command shortcuts: cognitive benefits and toolkit support. In: Proceedings of the 27th International Conference on Human Factors in Computing Systems (CHI 2009), pp. 2289–2298. ACM Press, New York (2009)
2. Ballagas, R., Rohs, M., Sheridan, J.G.: Sweep and point and shoot: phonecam-based interactions for large public displays. In: CHI 2005 extended abstracts, pp. 1200–1203. ACM Press, New York (2005)
3. Boring, S., Jurmu, M., Butz, A.: Scroll, tilt or move it: using mobile phones to continuously control pointers on large public displays. In: Proceedings of the 21st Annual Conference of the Australian CHI Special interest Group, OZCHI 2009, vol. 411, pp. 161–168. ACM, New York (2009)
4. Dachselt, R., Buchholz, R.: Natural throw and tilt interaction between mobile phones and distant displays. In: Proceedings of the 27th international conference extended abstracts on Human factors in computing systems (CHI 2009), pp. 3253–3258. ACM, New York (2009)

5. Duggan, M.: Wii Game Creation for Teens, 1st edn. Course Technology Press, Boston (2010)
6. Jorda, S., Hunter, S.E., i Conesa, P.P., Gallardo, D., Leithinger, D., Kaufman, H., Julia, C.F., Kaltenbrunner, M.: Development strategies for tangible interaction on horizontal surfaces. In: Proceedings of the Fourth International Conference on Tangible, Embedded, and Embodied Interaction (TEI 2010), pp. 369–372. ACM, New York (2010)
7. Kray, C., Nesbitt, D., Dawson, J., Rohs, M.: User-defined gestures for connecting mobile phones, public displays, and tabletops. In: Proceedings of MobileHCI 2010, pp. 239–248. ACM, New York (2010)
8. Malik, S., Ranjan, A., Balakrishnan, R.: Interacting with large displays from a distance with vision-tracked multi-finger gestural input. In: Proceedings of UIST 2005, pp. 43–52. ACM, New York (2005)
9. Moeslund, T.B., Hilton, A., Krüger, V.: A survey of advances in vision-based human motion capture and analysis. Computer Vision and Image Understanding 104(2), 90–126 (2006)
10. Nakamura, T., Takahashi, S., Tanaka, J.: Double-Crossing: A New Interaction Technique for Hand Gesture Interfaces. In: Lee, S., Choo, H., Ha, S., Shin, I.C. (eds.) APCHI 2008. LNCS, vol. 5068, pp. 292–300. Springer, Heidelberg (2008)
11. Nielsen, M., Störring, M., Moeslund, T.B., Granum, E.: A Procedure for Developing Intuitive and Ergonomic Gesture Interfaces for HCI. In: Camurri, A., Volpe, G. (eds.) GW 2003. LNCS (LNAI), vol. 2915, pp. 409–420. Springer, Heidelberg (2004)
12. Patel, S.N., Pierce, J.S., Abowd, G.D.: A gesture-based authentication scheme for untrusted public terminals. In: Proceedings of UIST 2004, pp. 157–160. ACM, New York (2004)
13. Paulson, B., Cummings, D., Hammond, T.: Object interaction detection using hand posture cues in an office setting. International Journal of Human-Computer Studies (2010), in press, accepted manuscript, Available online (October 14, 2010) doi: 10.1016/j.ijhcs.2010.09.003
14. Poppe, R.: Vision-based human motion analysis: An overview. Computer Vision and Image Understanding 108(1-2), 4–18 (2007)
15. Rico, J., Brewster, S.: Usable gestures for mobile interfaces: evaluating social acceptability. In: Proceedings of CHI 2010, pp. 887–896. ACM, New York (2010)
16. Shoemaker, G., Tang, A., Booth, K.S.: Shadow reaching: a new perspective on interaction for large displays. In: Proceedings of UIST 2007, pp. 53–56. ACM, New York (2007)
17. Vatavu, R.D.: Understanding challenges in designing interactions for the age of ambient media. In: Proceedings of the 3rd International Workshop on Semantic Ambient Media Experience (SAME 2010), Malaga, Spain, pp. 8–13. Tampere University Press (2010)
18. Wang, R.Y., Popović, J.: Real-time hand-tracking with a color glove. ACM Transactions on Graphics 28(3) (2009)
19. Vogel, D., Balakrishnan, R.: Distant freehand pointing and clicking on very large, high resolution displays. In: Proceedings of the 18th Annual ACM Symposium on User Interface Software and Technology, pp. 33–42. ACM, New York (2005)
20. Wobbrock, J.O., Morris, M.R., Wilson, A.D.: User-defined gestures for surface computing. In: Proceedings of the 27th International Conference on Human Factors in Computing Systems (CHI 2009), pp. 1083–1092. ACM, New York (2009)
21. Zhai, S., Kristensson, P.O., Gong, P., Greiner, M., Peng, S.A., Liu, L.M., Dunnigan, A.: Shapewriter on the iphone: from the laboratory to the real world. In: Proceedings of CHI 2009, pp. 2667–2670. ACM, New York (2009)

Graphs and Patterns for Context-Awareness

Andrei Olaru, Adina Magda Florea, and Amal El Fallah Seghrouchni

Abstract. A central issue in the domain of Ambient Intelligence is context - awareness. While previous research in the field presents complex context-aware infrastructures, but with little flexibility and fixed context representations, this paper presents a simple, flexible and decentralized representation of context, for the detection of appropriate context-aware action. This representation is inspired from notions like concept maps and conceptual graphs. A formalism for context patterns, that allows the detection and solution of problems, based on the user's context, is also proposed.

Keywords: Multi-agent system, context-awareness, ambient intelligence.

1 Introduction

A true Ambient Intelligence [5] system must be non-intrusive, but also proactive. Appropriate proactive action must fit the context of the user or the user will not have an optimal experience with the system [12]. Additionally, in order to be useful, many times proactive action must result from the anticipation of future situations,

Andrei Olaru
Computer Science Department, University Politehnica of Bucharest,
313 Splaiul Independentei, 060042 Bucharest, Romania,
and University Pierre et Marie Curie, 4 Place Jussieu, 75005 Paris, France
e-mail: `cs@andreiolaru.ro`

Adina Magda Florea
Computer Science Department, University Politehnica of Bucharest,
313 Splaiul Independentei, 060042 Bucharest, Romania
e-mail: `adina@cs.pub.ro`

Amal El Fallah Seghrouchni
Laboratoire d'Informatique de Paris 6, University Pierre et Marie Curie,
4 Place Jussieu, 75005 Paris, France
e-mail: `amal.elfallah@lip6.fr`

P. Novais et al. (Eds.): Ambient Intelligence - Software and Applications, AISC 92, pp. 165–172.
springerlink.com

based on the current context. Context-aware action and anticipation are key features that will make an AmI system seem "intelligent".

Context has been defined as: "Any information that can be used to characterize the situation of entities (i.e. a person, a place or an object) that are considered relevant to the interaction between a user and an application, including the user and the application themselves" [3]. Most works deal with context as location, location and time or other physical conditions (e.g. temperature) [1, 11], and some also with activity [6]. Situation may be described by means of associations [6], ontologies or rules [11], that are, however, predefined aspects.

In our approach towards an implementation of Ambient Intelligence [9, 10], we are trying to build mechanisms and representations that facilitate a more flexible approach to AmI and context-awareness, while in the same time are easy to implement and can work on resource-constrained devices.

In this paper we present a formalism that allows agents in a multi-agent system, that have only local knowledge, to share and process context-related information and to solve problems by using *context matching* and *context patterns*. Each agent has a representation of the context of its assigned user, including models of other users' context. The focus of this paper is more on defining a manner of representation and problem solving, and less on the algorithms used for context matching or the protocol for the communication between agents.

Context matching is based on representing information about the context of a user as a conceptual graph [13] and on the existence of context patterns, or, in short, *patterns*. Patterns can describe (in variable detail) situations that the user has experienced or common sense knowledge. Two matching context graphs (for two different users) mean a shared context, which is an occasion for further sharing of information between the users' agents. A pattern that matches the user's context means the user has been in the situation before (or it is a well-known, common sense, situation) and this allows the agent to anticipate possible problems, as well as to find solutions to problems already detected.

The next section presents related work in the field of context-awareness. The proposed context representation and scenario are presented in Section 3. Section 4 defines context matching and problem solving. The last section draws the conclusions.

2 Related Work

In previous work in the field of context-awareness there are usually two points of focus: one is the architecture for capturing context information; the other is the modeling of context information and how to reason about it.

Ever since the first works on context-awareness for pervasive computing [4], certain infrastructures for the processing of context information have been proposed [1, 6, 7]. There are several layers that are usually proposed, going from sensors to the preprocessing of the percieved information, the layer for its storage and

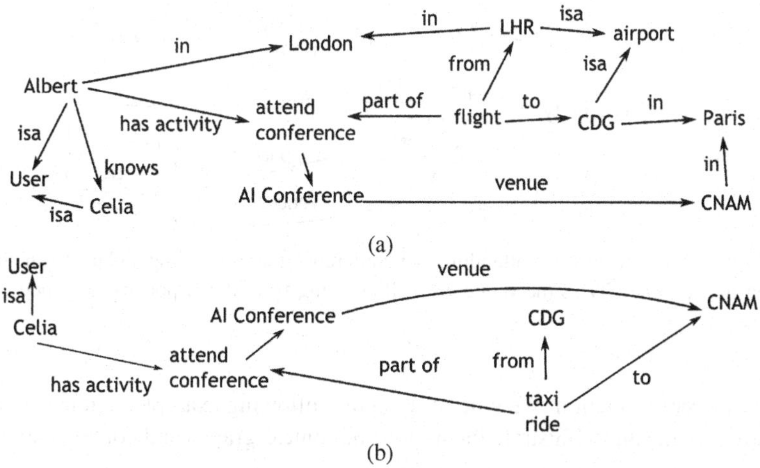

Fig. 1 The knowledge of Albert's and Celia's agents, respectively: (a) Albert will attend a conference in Paris; he will reach Paris by plane; (b) Celia will be attending the same conference, and will get from the CDG airport to the conference venue by taxi (her flight is omitted in this picture)

management, and finally to the application that uses the context information [1]. This is useful when the context information comes from the environment. However, physical context is only one aspect of context [2].

Infrastructures are usually centralized, using context servers that are queried to obtain relevant or useful context information [4, 7]. In our approach [9], we attempt to build an agent-based infrastructure that is decentralized, in which each agent has knowledge about the context of its user, and the main aspect of context-awareness is based on associations between different pieces of context information.

Modeling of context information uses representations that range from tuples to logical, case-based and ontological representations [11]. Henricksen et al use several types of associations as well as rule-based reasoning to take context-aware decisions [6]. While ontologies make an excellent tool of representing known concepts, context is many times just a set of associations that changes incessantly, so it is very hard to dynamically maintain an ontology that describes the user's context by means of concepts. In this paper we propose a more simple, but flexible and easy-to-adapt dynamical representation of context information, based on the notions of concept map and conceptual graph [8, 13]. While there has been a significant body of work in the domain of ontology alignment, this is not the subject of this paper. We assume ontologies have already been aligned.

3 Context Modeling

This goal of this paper is to present a formalism for the representation of context information, that can be used for assisting the user in solving problems. This section

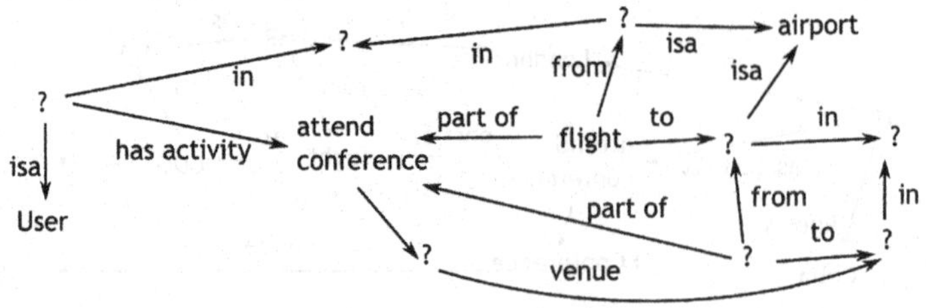

Fig. 2 A pattern that says that attending to a conference requires taking a plane to an airport located in the same city as the venue, as well as going from the airport to the venue of the conference

covers the representation. We will work on the following example, that is very simple and only meant to illustrate the notions of context graphs and context patterns:

Example. Albert is a researcher in the field of Artificial Intelligence. His professional agenda contains, among other activities, attending the AI Conference, which is held in Paris, at the CNAM. Albert has already booked a flight from Heathrow airport to Charles de Gaulle airport, but has not yet thought about how to get from the airport to CNAM. Celia, another researcher in the same field, will also attend the conference. Besides booking a flight, she has also noted in her agenda that she will be taking a taxi from the airport to the venue of the conference (this reminder will help her later be prepared with money to pay for the taxi). Albert and Celia know each other, and the communication between their respective agents will help solve Albert's problem.

Albert and Celia are both users of the AmIciTy Ambient Intelligence system. What we want is that the system (1) detects the need for a means of transportation for Albert and (2) based on information on Celia's agenda, suggest that a taxi may be an appropriate solution for Albert as well.

Each user of AmIciTy has an associated agent. Figure 1 shows the concept graph for the knowledge of the two agents (agent *A* for Albert and agent *B* or Celia) that is relevant to the scenario (the information about Celia's flight has been omitted in the figure). Formally, the knowledge of each agent can be represented as a graph $G = (V, E)$ in which the values of vertices and edges can be either strings or, better, URI identifiers that designate concepts, relations, people, etc. The value of an edge may be null.

The graph that an agent has contains the knowledge that the agent has about the user and about the user's context. The graph represents the context of the user, in the measure in which the agent has perceived it (or been informed of by the user or by another agent). The manner in which the context information has been gathered is not the focus of this paper.

For the detection of possible problems in the user's context, we define the notion of *context patterns*. A pattern represents a set of associations that has been observed to occur many times and that is likely to occur again. Patterns may come from past perceptions of the agent on the user's context or be extracted by means of data mining techniques from the user's history of contexts. Commonsense patterns may come from public databases or be exchanged between agents. However, the creation or mining of patterns is not the subject of this paper. An example of a context pattern is presented in Figure 2.

Example. Albert is a researcher for some time now, so he has attended many conferences. His AmIciTy agent (agent A) has formed the following pattern: attending a conference located in a certain city usually implies a flight to that city. The flight is between two airports, so a means of transportation between the airport and the venue of the conference is also required.

A pattern is also a graph, but there are several additional features that makes it match a wider range of situations. The graph for a pattern s is defined as:
$G_s^P = (V_s^P, E_s^P)$
$V_s^P = \{v_i\}, v_i = ? \mid string \mid URI, i = \overline{1,n}$
$E_s^P = \{e_k\}, e_k = (v_i, v_j, E\_RegExp), v_i, v_j \in V_s^P, k = \overline{1,m}$, where E_RegExp is a regular expression formed of strings or URIs.

4 Context Matching

An AmIciTy agent has a set of patterns that it matches against the current context (graph G). A pattern G_s^P (we will mark with "^{P}" graphs that contain special features like ? nodes) *matches* a subgraph G' of G, with $G' = (V', E')$ and $G_s^P = (V_s^P, E_s^P)$, iff there exists an injective function $f : V_s^P \to V'$, so that

(1) $\forall v_i^P \in V_s^P, v_i^P = ?$ or $v_i^P = f(v_i^P)$ (*same value*)
and
(2) $\forall e^P \in E_s^P, e^P = (v_i^P, v_j^P, value)$ we have:
-if *value* is a string or an URI, then the edge $(f(v_i^P), f(v_j^P), value) \in E'$
-if *value* is a regular expression, then it matches the values $value_0$, $value_1, \ldots$, $value_p$ of a series of edges $e_0, e_1, \ldots, e_p \in E'$, where $e_0 = (f(v_i^P), v_{a_0}, value_0)$, $e_k = (v_{a_{k-1}}, v_{a_k}, value_1) k = \overline{1, p-1}$, $e_p = (v_{a_{p-1}}, f(v_j^P), value_p)$, $v_{a_l} \in V'$.

That is, every non-? vertex from the pattern must match a different vertex from G'; every non-RegExp edge from the pattern must match an edge from G'; and every RegExp edge from the pattern must match a series of edges from G'. Subgraph G' should be minimal.

A pattern G_s^P *k-matches* (matches except for k edges) a subgraph G' of G, if condition (2) above is fulfilled for $m - k$ edges in E_s^P, $k \in [1, m-1]$, $m = ||E_s^P||$ and G' remains connected and minimal.

Example. For the pattern in Figure 2 and the example in Figure 1 (a), the pattern 2-matches the knowledge about Albert that his agent has: Albert is attending a conference and he has booked a flight to the city in which the conference takes place. The missing edges relate to the trip from the airport to the venue: the *part of* edge,

```
pattern G_s^P k-matches G →
    if k > 0
        put problem in the list
    if there is a problem and this can be a solution
        is the solution is certain / complete
            let user know
            user confirms solution
                increase confidence in pattern
            otherwise
                decrease confidence in pattern
        otherwise
            link possible solution to problem
```

Fig. 3 Pseudo-code of the agent's behaviour related to context matching

the *from* edge and the *to* edge, and there is also the means of transportation that is missing. Since something is missing, the agent should ask Albert about that piece of information or to try to find it itself.

This is defined as a *problem*. A *Problem* is a graph G^P that is a partial instantiation of a pattern G_s^P, according to the current context. It is the union between (1) the subgraph G' of G that *k-matches* pattern G_s^P and (2) the part of G_s^P that is not matched by G' (the *unsolved* part of the problem). The problem remains associated with the pattern: for the pattern $G_s^P = (V_s^P, E_s^P)$ and the k-matching subgraph $G' = (V', E')$, the problem p is a tuple (G_s^P, G_p^P), where G_p^P is the problem's graph:

$G_p^P = G' \cup G_x^P$, $G_x^P = (V_x^P, E_x^P)$ *(the unsolved part)*, where
$V_x^P = \{v \in V_s^P, v \notin dom(f)\}$
$E_x^P = \{e \in E_s^P \textit{ for which condition } (2) \textit{ is not fulfilled}\}$
G_x^P is a subgraph of G_s^P.

Example. In our scenario, the matching part of the problem contains the nodes *Albert*, *User*, *London*, *attends conference*, *AI Conference*, *flight*, *LHR*, *CDG*, *airport*, *from*, *to*, *Paris*, *CNAM*, and their connecting edges. The unsolved part of the problem contains one ? node and the edges *part of*, *from* and *to*.

But the agents are part of the AmIciTy multi-agent system. Agents *A* and *B* communicate because Albert and Celia know each other (they share common social context). They also share the same field of research. Based on the fact that the node *AI Conference* relates to Albert and to Artificial Intelligence as a domain (the same nodes to which Celia relates to), agent *A* will send to Celia the association between Albert, the domain, and the conference. This is common context, so Celia will respond with the data connected to the conference: venue and the means of transportation that she will use to get there (see Figure 1 (b)).

All the data that agent *A* has about other agents (here, agent *B*) is stored in the agent's knowledge base as its model of the other users. The model for the other users is not necessarily separate though: if the same concept appears in both models (provided the concept ha the same URI, or the agent is able to detect by means of common sense knowledge that it is the same concept), both subgraphs will contain the corresponding node. The model for Celia's agenda contains the same *AI Conference* node that is contained in the graph for Albert. When matching patterns from its pattern set, *A* detects that the pattern mentioned above fully matches the model for Celia. Agent *A* also has a *problem* that is linked to this pattern. Since Celia's context fully matches the pattern, it means it may be a *solution* to Albert's problem: Albert may also use a taxi to reach the conference.

In this particular case, with the given knowledge and patterns, there is only one solution to the problem that arose. But in a more realistic case, where context is more complex and there are more patterns, more solutions to the same problem may be found. In case they fit equally well in the current context, then the agent must prompt the user with all of them and the user must be given the choice.

It can be argued that context-matching is a very difficult problem in the case of large graphs and complex situations. However, resource-constrained devices will work only with smaller pieces of context information (i.e. smaller graphs), that are relevant to their function. Second, algorithms inspired from data-mining allow for incremental matching, starting from common nodes and growing the matching sub-graph.

Another problem that may appear in realistic situations (as opposed to our simple example) is the abundance of simultaneous matching context patterns, possibly describing contradictory situations. This is where more refined measures must be found that will allow calculating the relevance of each match. This too will be part of our future work.

5 Conclusion

So far, work in context-awareness for pervasive environments has been based predominantly on location-awareness and physical conditions. The use of ontologies or rules does not bring much dynamical flexibility and they are not easy to modify automatically, at runtime.

This paper proposes a more simple and more flexible manner of representing context and context patterns, that allows the agents to take decisions without the need for a centralized structure, by means of their knowledge, their history and by local communication alone.

At this point, we are in the process of identifying an efficient matching algorithm, as well as deploying the described formalism into a previously implemented multi-agent system. There is a great potential in detecting incompatible contexts – contexts that the user should not be in. Also, uncertainty and temporal relations have yet to be included in our work.

Acknowledgment. This work was supported by CNCSIS - UEFISCSU, project number PNII - IDEI 1315/2008 and by the Sectoral Operational Programme Human Resources Development 2007-2013 of the Romanian Ministry of Labour, Family and Social Protection through the Financial Agreement POSDRU/6/1.5/S/16.

References

1. Baldauf, M., Dustdar, S., Rosenberg, F.: A survey on context-aware systems. International Journal of Ad Hoc and Ubiquitous Computing 2(4), 263–277 (2007)
2. Chen, G., Kotz, D.: A survey of context-aware mobile computing research. Technical Report TR2000-381, Dartmouth College (2000)
3. Dey, A.: Understanding and using context. Personal and Ubiquitous Computing 5(1), 4–7 (2001)
4. Dey, A., Abowd, G., Salber, D.: A context-based infrastructure for smart environments. In: Proceedings of the 1st International Workshop on Managing Interactions in Smart Environments (MANSE 1999), pp. 114–128 (1999)
5. Ducatel, K., Bogdanowicz, M., Scapolo, F., Leijten, J., Burgelman, J.: Scenarios for ambient intelligence in 2010. Tech. rep., Office for Official Publications of the European Communities (2001)
6. Henricksen, K., Indulska, J.: Developing context-aware pervasive computing applications: Models and approach. Pervasive and Mobile Computing 2(1), 37–64 (2006)
7. Lech, T.C., Wienhofen, L.W.M.: AmbieAgents: a scalable infrastructure for mobile and context-aware information services. In: Proceedings of the 4th International Joint Conference on Autonomous Agents and Multiagent Systems (AAMAS 2005), Utrecht, The Netherlands, July 25-29, pp. 625–631 (2005)
8. Novak, J.D., Cañas, A.J.: The origins of the concept mapping tool and the continuing evolution of the tool. Information Visualization 5(3), 175–184 (2006)
9. Olaru, A., El Fallah Seghrouchni, A., Florea, A.M.: Ambient intelligence: From scenario analysis towards a bottom-up design. In: Essaaidi, M., Malgeri, M., Badica, C. (eds.) Proceedings of IDC 2010, the 4th International Symposium on Intelligent Distributed Computing., vol. 315, pp. 165–170. Springer, Heidelberg (2010)
10. Olaru, A., Gratie, C.: Agent-based information sharing for ambient intelligence. In: Essaaidi, M., Malgeri, M., Badica, C. (eds.) Proceedings of IDC 2010, the 4th International Symposium on Intelligent Distributed Computing, MASTS 2010, the The 2nd International Workshop on Multi-Agent Systems Technology and Semantics. SCI, vol. 315, pp. 285–294. Springer, Heidelberg (2010)
11. Perttunen, M., Riekki, J., Lassila, O.: Context representation and reasoning in pervasive computing: a review. International Journal of Multimedia and Ubiquitous Engineering 4(4), 1–28 (2009)
12. Riva, G., Vatalaro, F., Davide, F., Alcañiz, M. (eds.): Ambient Intelligence. IOS Press, Amsterdam (2005)
13. Sowa, J.: Knowledge representation: logical, philosophical, and computational foundations. MIT Press, Cambridge (2000)

ARTIZT: Applying Ambient Intelligence to a Museum Guide Scenario

Oscar García, Ricardo S. Alonso, Fabio Guevara, David Sancho,
Miguel Sánchez, and Javier Bajo

Abstract. Museum guides present a great opportunity where the Ambient Intelligence (AmI) paradigm can be successfully applied. Together with pervasive computing, context and location awareness are the AmI features that allow users to receive customized information in a transparent way. In this sense, Real-Time Locating Systems (RTLS) can improve context-awareness in AmI-based systems. This paper presents ARTIZT, an innovative AmI-based museum guide system where a novel RTLS based on the ZigBee protocol provides highly precise users' position information. Thus, it can be customized the content offered to the users without their explicit interaction, as well as the granularity level provided by the system.

Keywords: Ambient Intelligence, museum guide, context-aware, location-aware, Real Time Locating Systems, ZigBee.

Oscar García
School of Telecommunications, University of Valladolid, Paseo de Belén 15,
47011 Valladolid, Spain
e-mail: oscgar@tel.uva.es

Ricardo S. Alonso
Department of Computer Science and Automatic, University of Salamanca,
Plaza de la Merced, s/n, 37008, Salamanca, Spain
e-mail: ralorin@usal.es

Fabio Guevara · David Sancho
Nebusens, S.L. R&D Department, Parque Científico de la USAL, Edificio M2,
Calle Adaja s/n, 37185, Villamayor, Salamanca, Spain
e-mail: {fabio.guevara,david.sancho}@nebusens.com

Miguel Sánchez · Javier Bajo
Pontifical University of Salamanca, C/ Compañía 5, 37002, Salamanca, Spain
e-mail: jbajope@upsa.es

P. Novais et al. (Eds.): Ambient Intelligence - Software and Applications, AISC 92, pp. 173–180.
springerlink.com

1 Introduction

Ambient Intelligence (AmI) systems have to take into consideration the context in which they are used. That is, they must have context-awareness properties and adapt their behavior without the need of users to make an explicit decision or interact with them, allowing applications to be more usable and efficient [1]. People, places and objects are recognized as the three main entities when dealing with Ambient Intelligence [2]. The place where the user is and the objects that surround him determine the behavior of the system, thus obtaining in a natural way personalized, adaptive and immersive applications.

Mobile devices, such as smart phones, PDAs or tablets, offer a wide range of possibilities to create new AmI-based systems. One important feature of these devices is the ability to know their position, which includes the location of users themselves and any other object that is part of the environment [2]. However, the use of these devices on context-aware applications requires locating them more precisely. In this sense, Real-Time Locating Systems (RTLS) acquire a great importance in order to improve the applications based on the knowledge of the relative position of each user or object at any time.

One of the areas of interest where AmI becomes more relevant is Museum Guides applications [3]. Museum scenarios are environments where users receive a wealth of information from many sources. New information and communication technologies facilitate that the characteristics and related information about artworks can be offered in a more understandable, attractive and easy way to visitors. In this sense, the context information becomes relevant in order to personalize the experience for every user at every moment [4]. Thus, RTLS are presented as a resource that greatly improves context-awareness in AmI applications as these systems provide the position of every static or dynamic object that interacts throughout the scenario. There are different technologies that can be used when designing and deploying an RTLS, such as Global Positioning System (GPS) [5], Infrared (IR) Pointing Systems [6], Passive and Active Radio-Frequency Identification (RFID) [7], Wireless Local Area Networks (WLANs) [8] or Near Field Communication (NFC) [9].

There are several works that use RTLS to enhance the visitors experience in museums [8,10]. The ultimate goal of all these approaches is to find an association between the visitors and the art-work they are looking at a particular time. However, most of these solutions require a strong interaction between the visitors and the technology. That is, visitors must be close enough to the art-work to be detected, even bringing their devices to a particular object or waiting for other visitors to finish their interaction. Furthermore, these approaches do not provide enough accuracy to customize information with a high granularity level.

This paper describes ARTIZT, an innovative museum guide system in which visitors use tablets to receive personalized information and interact with the rest of the elements in the environment. An RTLS based on ZigBee is proposed in order to improve context-awareness. This system achieves a location precision of less than one meter. Thus, ARTIZT knows at every time how customized content must be shown to the visitors, based on the area on which they are located. This way, visitors can naturally walk throughout the museum receiving relevant information as they get closer to each art-work. With this approach, every art-work, zone or important detail of the museum can be contextualized.

The rest of the paper is structured as follows: Section 2 presents existing museum guide works, as well as the shortcomings that motivate the development of a new system. Then, Section 3 describes the basic components of ARTIZT, including the most important locating and context management technologies involved on it and how they are used. Finally, conclusions and future work are depicted.

2 Problem Description and Related Work

In recent years, ongoing advances in technology and communications have allowed people to be surrounded by mobile devices. These devices are increasingly powerful, easy-to-use and capable of communicating with each other in more innovative ways. These capabilities allow the creation of Ambient Intelligence systems: users' mobility is guaranteed and they are, day by day, more accustomed to the use of technology for any task.

A museum scenario is an ideal environment where AmI can improve the way the information is offered to visitors. AmI features encourage the creation of new museums guide approaches where information is presented in a natural, personalized and attractive way to users with a better human-machine interaction. This way, as art-works are usually statically placed in the environment, the main goal is how to detect where visitors are. If we can determine, as precisely as possible, the position of the visitor inside the museum, we will be able to know the entire context that surrounds him every time.

Multiple technologies can be used in order to determine the position of the visitors. The ultimate goal of knowing the positioning of the visitors is to provide information, as accurate as possible, about the art-works they are watching. In this sense, tagging the context and the use of RTLS are two widely used alternatives [11]. Next, it is analyzed the most used technologies in both trends.

When talking about RTLS, the first technology that comes into our mind is GPS. It has been used in context-aware tourist guides where users receive personalized information according to their position [12]. However, it only works outdoors and it is not an appropriate technology to develop an indoor museum guide. Some approaches try to solve indoor locating problem using combined technologies along GPS. Cyberguide [5] determines the users' position by means of infrared sensors. However, it is needed a direct line of sight between user and sensors, so it does not work properly in crowded environments. Exploring the use of infrared sensors, the HIPPIE system delivers information about the location of user in relation to an object [6]. Visitors point the art-work (which is provided with an infrared detector) and the system knows where they are and provides the content. Nevertheless, this solution requires a proactive user and, as mentioned before, direct line of sight and proximity between user and object. The action performed by the user when pointing the object (*e.g.*, an art-work) is physically the same as is done in solutions that use RFID [7] or NFC technologies [9]. Both technologies follow the same pattern of performance: each object all over the museum is tagged. RFID or NFC tags, containing a unique identification number, are placed near the object. If visitors want to receive information about an art-work, they have to place their device nearby the tag. Then, the system identifies the object and loads the relevant information. The necessity of proximity between devices and tags makes these solutions non-transparent to the user as it requires a direct collaboration. Moreover, if multiple users want to see the information about the same art-work, they must do it one by one at each time. Bluetooth solves the peer-to-peer relation between RFID tags and users [14]. The Eghemon system uses mobile phones that receive information via Bluetooth as users get closer to a piece of information (*e.g.*, an art-work). These pieces can offer information to multiple devices simultaneously, but they must be close enough to them. If two pieces are close enough, the mobile device can only receive information from just one of them. Bluetooth can also be used to improve the devices locating. Bruns *et al.* have designed a solution where a grid of Bluetooth emitters is deployed throughout the museum, so the visitors' mobile phones transmit their position to the nearest emitter [10]. However, location is not as accurate as desirable, because the system only knows which mobile phones are associated to an emitter. Furthermore, Bluetooth can only support the association of up to seven mobile phones simultaneously to an emitter. The UbiCicero system [14] uses Active RFID to get the position of the users. In this case, users carry an RFID reader that continuously reads signals from active tags that are close to art-works. Active RFID technology allows a better locating approach because users do not have to be as close as with passive RFID technology to detect the object. However, location information is not precise enough because the system only detects which art-work is closer the users, so two objects that are close enough can cause interference and "confuse" the system. It is also possible to locate mobile devices that provide information on the museum via Wireless LANs. In this case, multiple Wireless LANs are created, usually one network for each different zone. When visitors change from a zone to another, devices automatically connect to an available network through which contents are provided [8]. This solution presents several problems: different coverage areas must not be overlapped; locating provides a poor precision (never gets precision under 2 meters); and infrastructure deployment is a hard process since it is needed a thorough calibration of all devices [13].

As can be seen along this section, there are many approaches that have been considered to create museum guides. Analyzing all of them, two ways of tackling the problem can be identified. The first one is the direct physical interaction between the visitor and the art-work. The second one consists of obtaining information without having a voluntary interaction by the visitor, where the system automatically provides information depending on visitors' location. ARTIZT follows the second approach. The main objective is to make interaction as transparent as possible to visitors. For doing that, it is very important that context information is precise enough. This way, an accurate location of visitors is the most important aspect of this approach. Therefore, it is easy to know which elements (*i.e.*, art-work) surround each visitor, no matter several of these elements are separated by a short distance between them. Thus, it is possible to get a whole description of the context that surrounds each visitor in a more precise way. Based on the context information, the content provided to the visitors changes dynamically. Next section describes the basic features of ARTIZT and the way it works.

3 ARTIZT: Ambient Intelligence Real-Time Locating System Museum GuIde Over Zigbee Technology

The AmI paradigm proposes the development of applications that provide new ways of interaction between people and technology, adapting them to the needs of individuals and their environment [2]. ARTIZT (*Ambient intelligence Real-Time locating system museum guIde over Zigbee Technology*) follows these premises and offers personalized contents to the visitors in a transparent way according to the context information.

The potential of ARTIZT lies in the precision with which contextual information can be collected. In this sense, an innovative RTLS provides the system with users' positions with an error less than one meter, so it can be determined at any time which art-works are on the visitors' radio of interest, thus adapting precisely the information that it is provided to each user.

3.1 The Real-Time Locating System

The RTLS used is by ARTIZT is based on the novel n-Core platform [15], which is intended to develop ZigBee applications and provides both wireless physical devices and an Application Programming Interface (API) to access their functionalities. A network of ZigBee devices must be deployed all over the museum. This network is composed of a set of *Sirius A* (Figure 1 left) and *Sirius Dongle* devices (Figure 1 right). The *Sirius A* devices are placed across the ceiling of the museum (Figure 2), forming a network in which it is known the specific location of each one, as well as the relative positions with each of its neighbors (*i.e.*, closest devices). The Sirius Dongle devices are inserted in tablet PCs that are carried by the visitors. This way, the visitors can move freely through the museum.

Fig. 1 *Sirius A* (left) and *Sirius Dongle* (right) devices

All devices communicate via the ZigBee standard. ZigBee is a low cost, low power consumption wireless communication standard, developed by the ZigBee Alliance. It is based on the IEEE 802.15.4 protocol and operates at the

868/915MHz and 2.4GHz unlicensed bands. ZigBee is designed to be embedded in consumer electronics, home and building automation or toys and games. ZigBee allows star, tree or mesh topologies. Devices can be configured to act as network coordinator (creates and controls the network) router (sends/receives/forwards data to/from other devices) and end device (sends/receives data to/from other devices in the network).

Over this network infrastructure it is implemented a locating engine that provides users positioning whose accuracy reaches less than one meter. The infrastructure is completely dynamic and scalable so new devices can be added at any time without affecting the rest of the network. The operation of the RTLS is very simple. Visitors carrying mobile devices (*i.e.*, tablets) move freely around the museum. The mobile devices send periodically a broadcast signal by means of the *Sirius Dongle* connected to them. The signal is received by the *Sirius A* devices placed all over the museum. The location engine, allocated in a central server, calculates the positions of all *Sirius Dongle* devices and therefore the position of every visitor. Once the system knows the location of each visitor, an application installed on the devices of the visitors customizes the information which is provided dynamically.

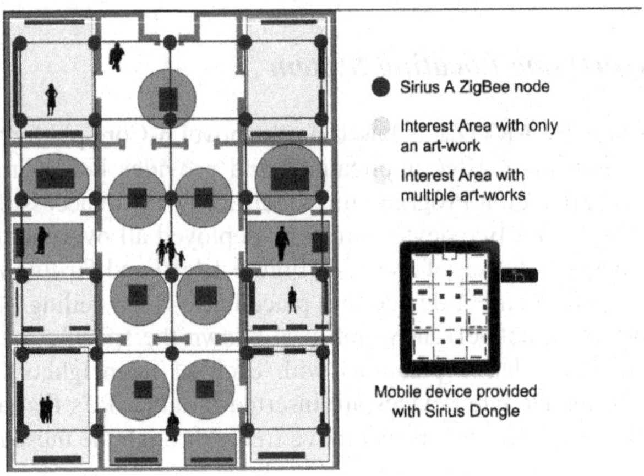

Fig. 2 *Sirius A* wireless network, *Interest Areas* and mobile device provided with Sirius Dongle

3.2 Context Information Management

Once ARTIZT gets the position of the visitors, the information must be personalized according to every visitor status. Each user carries a tablet on which it is installed a light and user-friendly application developed specifically for the museum. Contextual information of the museum is included in this application so that, from the position of the visitors, ARTIZT decides which information is shown in the device. To do this it is created a map of the museum and, with the location information received by the server, it is determined the location of visitors continuously. In addition, each level of the museum is divided into "*Interest Areas*" (see Figure 2). So when a user enters into one of these, the application customizes all the information that is wanted to be shown to visitors. Reader can realize that with a so precise RTLS, these areas can be as small as desired, so that the system can provide enhanced context information and personalize it always in a transparent way without any user interaction.

Tablets' application contains all the information that may be provided to users. Thus, ZigBee network only carries data from the devices to the server, necessary to calculate their position, and the location of the visitors from the server to the tablets. These communications reduce traffic data and there are always alternative paths in the network so fault tolerance increases and data loss is minimized.

ARTIZT is configured to provide the customized content. When the position of the visitor is known, and using a series of data that is collected before the visitor starts his route, the system is able to tailor the information that is being shown in the form of content or language.

4 Conclusions and Future Work

Transparency and customized data is one of the keys in the development of AmI applications. If these applications want to adapt their behavior depending on the context that surrounds users transparently, it is important that users' location is as precise as possible.

Museum guides are a scenario where applying AmI techniques becomes more meaningful: multiple contextual information that contains the museum and the great mobility of the visitors make their location and the data filtering process become challenges to solve in an efficient way.

ARTIZT is a museum guide system that includes AmI techniques. This system makes use of a powerful RTLS to determine where visitors are. This RTLS, based on an innovative ZigBee-based platform, provides high precision in the locating process. Therefore, the information that surrounds each visitor at any moment can be precisely determined and provided to them on real-time. The use of the ZigBee network to transport locate information minimize data loss (visitors' devices contains all the necessary information) and reduce the risk of disconnection, as there are always alternative ways to reach the server over the ZigBee grid.

Future work on ARTIZT includes the use of sensors in order to determine any information that may affect the context (temperature, pressure, humidity, etc.) and estimate the direction of art-works and visitors through the use of compasses. Combining the precise location information provided by the RTLS with the information from compasses it will be possible to determine where visitors are looking at, enhancing the user experience.

Acknowledgments. This work has been supported by the Spanish Ministry of Science and Innovation, Project T-Sensitive, TRA2009_0096.

References

1. Baldauf, M., Dustdar, S., Rosenberg, F.: A survey on context-aware system. International Journal of Ad Hoc and Ubiquitous Computing 1(4), 263–177 (2007)
2. Weber, W., Rabaey, J.M., Aarts, E.: Ambient Intelligence. Springer-Verlag New York, Inc., Heidelberg (2005)
3. Ramos, C., Augusto, J.C., Shapiro, D.: Ambient Intelligence—the Next Step for Artificial Intelligence. IEEE Intelligent Systems 23(2), 15–18 (2008)
4. Raptis, D., Tselios, N., Avouris, N.: Context-based design of mobile applications for museums: a survey of existing practices. In: Proceedings of the 7th International Conference on Human Computer Interaction with Mobile Devices & Services, Salzburg, Austria, September 19-22 (2005)
5. Abowd, D.A., Atkeson, C.G., Hong, J., Long, S., Pinkerton, M.: Cyberguide: a mobile context-aware tour guide. Wireless Networks 3(5), 421–433 (1996)
6. Oppermann, R., Specht, M.: A context-sensitive nomadic exhibition guide. In: Thomas, P., Gellersen, H.-W. (eds.) HUC 2000. LNCS, vol. 1927, pp. 127–142. Springer, Heidelberg (2000)
7. Bellotti, F., Berta, R., De Gloria, A., Margarone, M.: Guiding Visually Impaired People in the Exhibition. In: Mobile Guide 2006, Turin, Italy (2006)
8. Cheverst, K., Davies, N., Mitchell, K., Smith, P.: Providing tailored Context-aware information to city visitors. In: Brusilovsky, P., Stock, O., Strapparava, C. (eds.) AH 2000. LNCS, vol. 1892, pp. 73–85. Springer, Heidelberg (2000)
9. Blöckner, M., Danti, S., Forrai, J., Broll, G., De Luca, A.: Please touch the exhibits!: using NFC-based interaction for exploring a museum. In: Proceedings of the 11th International Conference on Human-Computer Interaction with Mobile Devices and Services (MobileHCI 2009), Bonn, Germaby, pp. 71–72. ACM, New York (2009)
10. Bruns, E., Brombach, B., Zeidler, T., Bimber, O.: Enabling Mobile Phones To Support Large-Scale Museum Guidance. IEEE Multimedia 14(2), 16–25 (2007)
11. Ghiani, G., Paterno, F., Santoro, C., Spano, L.D.: UbiCicero: A location-aware, multi-device museum guide. Interacting with Computers 21(4), 288–303 (2009)
12. Park, D.-J., Hwang, S.-H., Kim, A.-R., Chang, B.-M.: A Context-Aware Smart Tourist Guide Application for an Old Palace. In: International Conference on Convergence Information Technology, November 21-23, pp. 89–94 (2007)
13. Zimmermann, A., Lorenz, A.: LISTEN: a user-adaptive audio-augmented museum guide. User Modeling and User-Adapted Interaction 18(5), 389–416 (2008)
14. Bay, H., Fasel, B., Van Gool, L.: Interactive museum guide. In: The Seventh International Conference on Ubiquitous Computing UBICOMP, Workshop on Smart Environments and Their Applications to Cultural Heritage (2005)
15. n-Core: A Faster and Easier Way to Create Wireless Sensor Networks (2010), `http://www.n-core.info` (retrieved November 14, 2010)

Reliability of Location Detection in Intelligent Environments

Shumei Zhang, Paul J. McCullagh, Chris Nugent,
Huiru Zheng, and Norman Black

Abstract. Radio Frequency Identification (RFID) technology has been used in Intelligent Environments to track objects and people, but the technology is subject to reliability issues of sensor malfunction, sensor range, interference and location coverage. This paper discusses the optimal deployment for a fixed RFID reader network in an indoor environment with the aim of achieving more accurate location whist minimizing the equipment costs. Given that data may be occasionally lost, a rule-based pre-processing algorithm was developed for missing data judgment and correction, to improve the robustness of the technique. The algorithms were evaluated using experiments of single mobile tag and multiple mobile tags. The average subarea location accuracy based on pre-processed and original data is 77.1% vs. 68.7% for fine-grained coverage, and 95.8% vs. 85.4% for the coarse-grained coverage.

Keywords: Reliability, Location based systems, RFID, Missing data estimation.

1 Introduction

For an environment to be considered 'intelligent', sensors must be able to convey information regarding a person or an object to some 'supervisory' technology for assessment and sometimes intervention. In health and wellbeing research location determination is often a key factor, which can then be used to support ambient assisted living (AAL). Indoor location tracking systems use a variety of technologies, dependent upon accuracy, range and infrastructure costs. Wireless sensor networks (WSNs) [Mao 07] are low cost, small in size and straightforward to install. They have been used for environment monitoring, object tracking, health monitoring and hazard detection. Sensing technologies include Infra-Red

---

Shumei Zhang · Paul J. McCullagh · Chris Nugent · Huiru Zheng · Norman Black
University of Ulster, United Kingdom
email: {zhang-s2,pj.mccullagh,cd.nugent,h.zheng,
nd.black}@ulster.ac.uk

P. Novais et al. (Eds.): Ambient Intelligence - Software and Applications, AISC 92, pp. 181–188.
springerlink.com

(IR), ultrasound, IEEE 802.11 (Wi-Fi), Ultra Wide Band (UWB), and IEEE 802.15 (Bluetooth). A comparative study of WSNs [Ni 04] concluded that Radio Frequency Identification (RFID) is the most appropriate technology for indoor location sensing.

An RFID system comprises readers and tags, communicating wirelessly using radio waves of a defined frequency, using an agreed protocol. The RFID reader emits a challenge, and the tag responds by sending back its identity. The RFID tag is applied to or incorporated into an object or person. The communication range is a function of the reader design, the type of tag and the operational environment. RFID tags can be active, passive, or assisted passive (requiring an external source to 'wake up'). Active tags have a longer communication range, however, passive tags have a longer lifetime, are smaller and less expensive.

We investigate RFID deployment, and develop tracking algorithms with the aim of improving localization accuracy and reliability. In particular we use an approach to optimize the RFID reader network deployment. Due to the problems with intermittent data collection, we also use a preprocessing algorithm to impute missing values. The remainder of the paper is structured as follows. In Section 2, related work is discussed. Deployment of the RFID system used is described in Section 3. In Section 4, the missing data judgment and correction algorithms are explained. Section 5 provides our methodology for location classification experiments and presents the results of these experiments. Discussion and conclusions are provided in Section 6.

2 Related Work

Bahl et al. [Bahl 2000] presented an IEEE 802.11 wireless networking building-wide location tracking system called RADAR. Harter et al. [Harter 02] investigated location accuracy of the 'Active Bat', comprising two base stations and 100 ultrasound receivers in a laboratory environment. 95% of readings were computed within 9cm of their true positions. Nevertheless the approach was computationally expensive requiring up to 15 receivers for each position calculation. Both approaches would be impractical in an everyday AAL scenario. Minimizing the deployment and infrastructure costs whilst providing a precise indoor location tracking service remain an open research problem [Woodman 08].

RFID provides a promising solution offering information on both location and identity awareness. Application domains are diverse including retailing, aviation, healthcare, library services and construction [Ngai 08]. Satoh [Satoh 09] proposed a framework for location-aware communication in intelligent environments. The framework consisted of two components: a location detection model and a location-aware communication mechanism. Location computation methods use three categories of measurement: received signal strength (RSS), angle-of-arrival, and propagation time. The accuracy of location estimation using RSS suffers from errors caused by multi-path fading, temperature variations, furniture/door rearrangement, and mobile human presence [Naisbitt 10]. The received signal strength indicator (RSSI) is an approximation of radio signal strength as measured by the RF transceiver.

An RSS model is obtained by a priori measurements or in real-time by 'sniffing' devices deployed at known locations [Mao 07]. The model is subsequently used to match each presented signal strength vector to an estimated location using classification methods, which choose the location based on closest training examples. Elnahrawy et al. [Elnahrawy 04] had proposed three area-based algorithms: simple point matching which matches the RSS vector to a 'tile set' using thresholds; area-based probability which matches the RSS vector probabilistically; and Bayesian network which encodes the relationship between the RSS vector and the estimated location based on a signal-versus-distance propagation model. The three algorithms were compared and exhibited similar performance. Ni et al. [Ni 04] presented a location identification method based on a dynamic active RFID calibration (LANDMARC) approach, which employed fixed reference tags serving as reference points. Hallberg et al. [Halberg 09] proposed an approach for localization of forgotten items based upon LANDMARC and improved the accuracy and reduced number of reference tags.

3 Optimal Deployment Selection for RFID Reader Network

The implementation challenges in the RFID system deployment are determining how many readers need to be installed, and where their ideal placement is in the tracking environment. In this study, we used the Adage RFID products, Dyname Readers and Carme Active tags [Adage]. The measured readable range in our experiments was up to 4.5m with a robust range in the region of 4m. The experimental environment consists of a kitchen, and a living room [Nugent 09]. The combined size of the two rooms is 8.86m by 3.76m, with an active area of 7.8m by 2.75m (taking into consideration the furniture within the rooms). Deployment of readers was investigated by comparing the accuracy of location detection using different classifiers. The deployment of RFID reader network, according to the accuracy of location detection and equipment cost and the tracking subareas using coarse-grained coverage and fine-grained methods are illustrated in Figure 1.

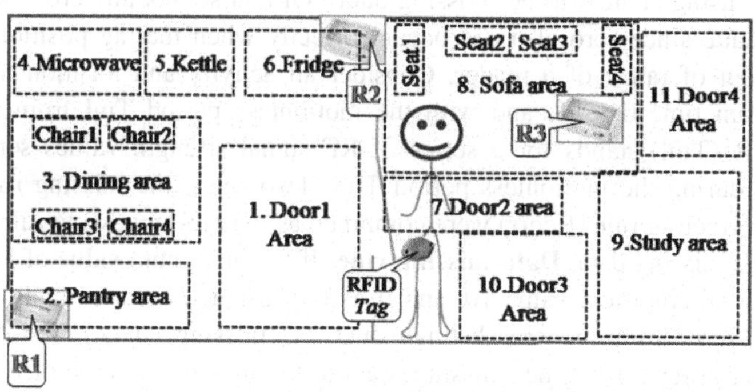

Fig. 1 Reader network deployment and functional subarea division using coarse-grained (11 subareas) fine-grained (17 subareas to include individual locations of seats and chairs) methods. R1, R2, and R3 are three readers.

The tracking environment can be divided into a number of functional subareas. In order to assess the relationship between the size of subareas and the accuracy of location detection, we defined two levels: fine-grained and coarse-grained. Using the coarse-grained categorization, 11 subareas were specified as shown in Figure1. Using the fine-grained categorization, the Dining area and Sofa area (area number 3 and 8) were further divided into four Chair areas and four Seat areas see Figure 1. In this case, a total of 17 fine-grained subareas were defined.

4 Rule-Based Algorithm for the RF Signal Pre-processing

Two common problems should be considered when using an RFID system: (1) Reader collision occurs in the area where the signals from two or more readers overlap, since the tag cannot respond to simultaneous queries; (2) Conversely, tag collisions occur when many tags are present in a small area [Kim 09]. Additionally, a problem with the RFID system was that the RF signal strength could be influenced by environmental factors such as: tracking of the tag carried by a person, absorption of the energy sent by the reader, or diminishing battery life in the tag [Hahnel 04]. In such a case the RF signal strength may vary and may even record zero erroneously within the normal operational radius between the tag and the reader. In data mining, we refer to such inconsistencies as 'missing data' [Darcy 10], which decreases the performance of the location detection algorithms.

In order to judge and correct the missing data, we developed a rule-based algorithm for pre-processing to improve the robustness of the location detection. In this algorithm the missing RF signal data can be judged according to the subject's activity postures. The three-dimensional accelerometer embedded in the HTC phone was used to measure the subject's acceleration for classification of their daily activities, such as walking, standing, sitting, lying or falling down [Zhang 09] [Zhang 10]. If the RF strength has a large fluctuation (e.g. recording an unexpected zero) when the subject is in an inactive period of time, we can judge this using context to be 'missing data'. Of course, not all zero instances are missing data, since zero also can occur correctly when the tag position is at the edge or out of range of a reader. Consider an activity and location monitoring period from time 0 to T, and with the motionless period Tml from tm to tl . Assume R (Tml) stands for a series of RF signal strength values sensed by a reader R during the motionless period Tml. Two rules: the missing rule (Rmis) and the correcting rule (Rcorr) were combined as the pre-processing rule (Rpre) to correct the missing data. **Data missing rule:** if the maximum value of R (Tml) is more than an empirical value 10, and if a zero instance occurs among R (Tml), then the zero instances are distinguished as 'missing' data, Eq. (1). **Data correcting rule:** correct the 'missing' data to the maximum value of R (Tml) as shown in Eq. (2).

$$R_{pre} = \begin{cases} \exists T_{ml} = [t_m , t_l] \in T; \quad \forall t_i \in T_{ml} \\ R_{mis:} \; R(t_i) = 0 \text{ and } \mathrm{Max}\{\mathrm{R}(t_m) : \mathrm{R}(t_l)\} \geq 10 & (1) \\ R_{corr:} \; R(t_i) = \mathrm{Max}\{\mathrm{R}(t_m) : \mathrm{R}(t_l)\} & (2) \end{cases}$$

An experiment was designed to demonstrate how this algorithm works for judging and correcting the missing RF signal strength according to the activity postures. Install one reader (R1) in the kitchen. A subject places an RFID tag in their pocket, a mobile holster on the left waist, and starts two programs at the same time. The subject enters through Door1 walks to the Fridge and stands for 1 minute, then walks to Chair2 and sits still for 1 minute on Chair2. The subject performs a gentle motion on Chair2 such as rotating 90 degrees to the left or right and maintains that posture for 1 minute. The subject then stops the two programs at the same time. The experiment was conducted with initially with one then with two readers (R1 and R2).

The experimental results for one reader R1 are shown in Figure 2 (a) and (b), and for two readers R1 and R2 in Figure 3 (a) and (b). The results presented in Figure 3 (a) and (b) show that the RF signal strength sensed by R1 was stable, and no missing data occurred. Nevertheless R2 was unstable. According to the pre-processing algorithm presented in Eq. (1) and Eq. (2), the 0 occurred at 39sec could be revised to 54 by the pre-processing algorithm. In the same way, the value 0 at the 198sec sensed by R2 could be corrected to 19 during the pre-processing stage.

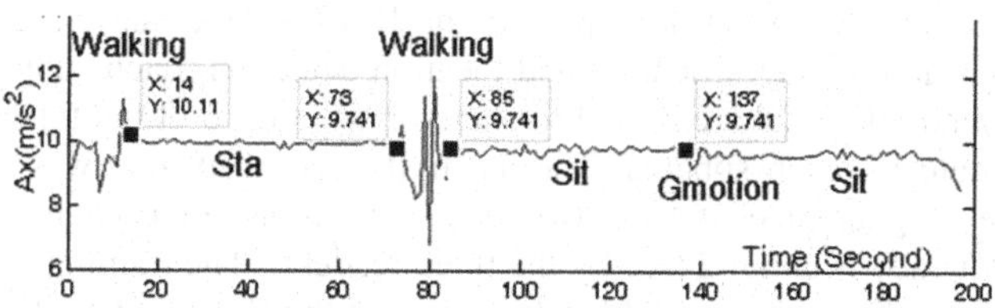

(a) Ax signal shows the subject's activity postures collected using the G-sensor.

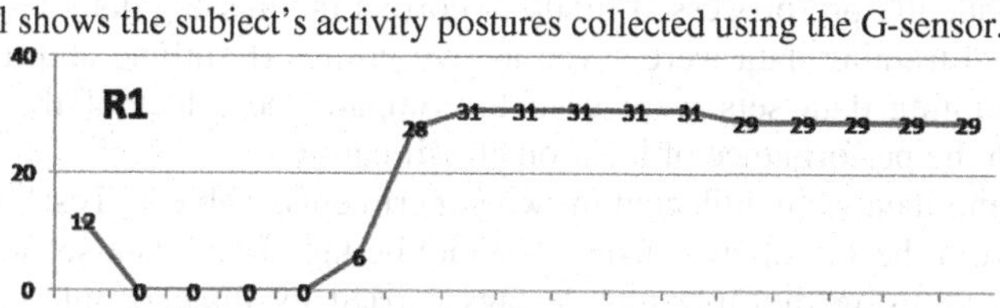

(b) RF signal strength values collected by R1 from a mobile tag.

Fig. 2 Comparison of the acceleration and RF signal strength for R1

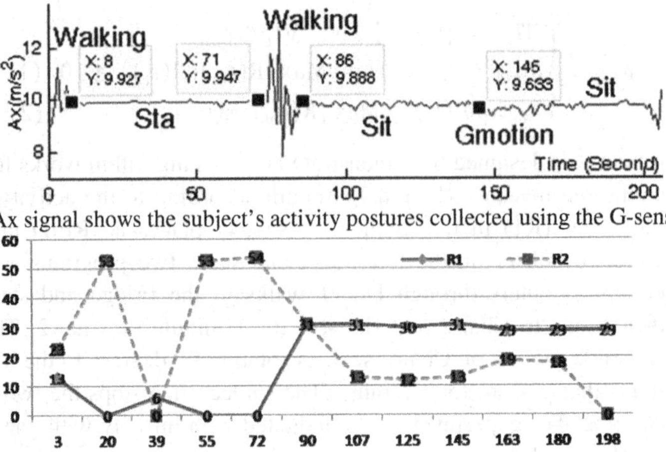

(a) Ax signal shows the subject's activity postures collected using the G-sensor.

(b) RF signal strength values collected by R1 and R2 respectively from a mobile tag

Fig. 3 Comparison of the acceleration and RF signal strength for readers R1 and R2

5 Smart Home-Based Experiments

Experimental results were compared: (1) fine-grained versus coarse-grained subareas; (2) pre-processed data set versus original data set. With the intention of imitating the natural dynamic activity of daily life, the tracking tag was carried by a person rather than using immobile reference tags. All subjects wore an HTC smart phone on the waist and placed an RFID active tag in a pocket for simultaneous activity classification and location detection. Each subjectwas asked to walk around all 17 subareas, and to stand or sit in each of the three boundaries for one minute for each subarea. The inactivity period of time was designed for one minute to guarantee at least two data repetitions for each position (RFID sensing frequency is about 24sec by the three readers). The training data collected were subsequently pre-processed firstly to correct the missing data. Following this the corrected training data were saved as two groups (trainFine and trainCoarse). The two training data sets were used to compare the effect of the size of the subareas on the performance of location classification.

The testing data were collected in two experiments, Table 1: Test1 data set was collected from the 17 subareas using one mobile tag; Test2 data set was collected from subareas using multiple mobile tags carried by three people concurrently. Each experiment was repeated three times by subjects following natural behavior i.e. there was no limitation on their position or the direction in which they faced, although it is known that their positions or rotations can influence the RF signal strength values. The data sets were classified using LibSVM (support vector machine) classifier [Chang 01] and compared from three aspects: (1) original vs. pre-processed; (2) single mobile tag vs. multiple mobile tags; (3) fine-grained vs. coarse-grained. The experimental results are presented in Table1. Accuracy improvement due to the pre-processing rules depends on the number of missing

Table 1 Testing data organization for evaluation of the pre-processing algorithm based on the coarse-grained and fine-grained subareas

Experiment	Subareas	Original Data Set	Pre-processed Data Set	Corresponding Training set
Test1	Coarse	Test1: 87.5%	CoarsePre: 91.67%	trainCoarse
		Test2: 83.3%	CoarsePre: 100%	
	Fine	Test1: 79.15%	FinePre: 79.17%	trainFine
Test2		Test2: 58.3%	FinePre: 75%	

data points. The accuracy was similar for the Test1 but improved for the Test2, as more missing data occurred during the Test2 experiment (multiple mobile tags). Additionally the coarse-grained method attains higher classification accuracy than the fine-grained method.

6 Discussion and Conclusion

This study proposed solutions to improve the location recognition accuracy and robustness for RFID systems. Initially, we discussed an efficient method for deployment of an RFID reader network based upon theory and experiment. An appropriate RFID deployment should minimize equipment costs whilst maintaining location accuracy. Then, a rule-base pre-processing algorithm was proposed to judge and correct missing data, which took account of user activity postures. The pre-processing algorithm was evaluated by the LibSVM classifier, based on two experiments *Test1* and *Test2* (single mobile tag and multiple mobile tags). The results demonstrated that this algorithm works well if the missing data occurred during the subject's period of inactivity. Correcting the missing data according to the synchronized sensing inactivity period of time can improve reliability. Hence location detection will be robust using the pre-processed data set, even though the original RF signal strength may be erratic. The use of multiple concurrent mobile tags around a small tracking area has the potential to cause more missing data. The average subarea location accuracy for *Test1* and *Test2* based on pre-processed and original data is 77.1% vs. 68.7% for the fine-grained method, and 95.8% vs. 85.4% for the coarse-grained methods.

At present the location and activity data are collected and saved in a computer and HTC phone respectively, and the data pre-processing and classification are carried out subsequent to the data collection. Hence location cannot be inferred in real time, which precludes interactive operation. However, it is possible to wirelessly communicate between the RFID reader and a smart phone and in future it would be possible to deploy algorithms onto a mobile device (e.g. a smart phone), which would be appropriate to an assisted living application, for example where an occupant's location is tracked for safety reasons. RF signal strength sensing frequency is decided by the reader and tag numbers. More readers or tags will increase the data sensing time, and the computational effort will increase proportionally. This is not a problem with the current computation but scalability would need careful consideration for an interactive solution Thus for an environment where the person moves quickly between locations, we would anticipate some undetected locations, which may need further investigation and correction by an enhanced algorithm.

References

1. Mao, G., Fidan, B., Anderson, B.: Wireless sensor network localization techniques. Computer Networks 51(10), 2529–2553 (2007)
2. Ni, L.M., Liu, Y., Lau, Y.C., Patil, A.P.: LANDMARC: indoor location sensing using active RFID. Wireless Networks 10(6), 701–710 (2004)
3. Bahl, P., Padmanabhan, V.: RADAR: An in-building RF-based user location and tracking system. IEEE Infocom, 775 (2000)
4. Harter, A., Hopper, A., Steggles, P., Ward, A., Webster, P.: The anatomy of a context-aware application. Wireless Networks 8(2), 187–197 (2002)
5. Woodman, O., Harle, R.: Pedestrian localisation for indoor environments. In: Proceedings of the 10th International Conference on Ubiquitous Computing, p. 114. ACM, New York (2008)
6. Ngai, E., Moon, K.K.L., Riggins, F.J., Yi, C.Y.: RFID research: An academic literature review (1995-2005) and future research directions. International Journal of Production Economics 112(2), 510–520 (2008)
7. Satoh, I.: Location-aware communications in smart environments. Information Systems Frontiers 11(5), 501–512 (2009)
8. Naisbitt, J.D.: Reliable location sensing through multi-sensor fusion, dynamic weighting, and confidence mapping, Thesis, University of Illinois at Urbana-Champaign (2010)
9. Elnahrawy, E., Li, X., Martin, R.P.: The limits of localization using signal strength: A comparative study. In: IEEE Conf. on Sensor and Ad Hoc Communications and Networks, pp. 406–414 (2004)
10. Hallberg, J., Nugent, C., Davies, R., Donnelly, M.: Localisation of Forgotten Items using RFID Technology. In: Proceedings of the 9th International Conference on Information Technology and Applications in Biomedicine, Larnaca, Cyprus (2009)
11. Nugent, C.D., Mulvenna, M.D., Hong, X., Devlin, S.: Experiences in the development of a smart lab. IJBET 2(4), 319–331 (2009)
12. ADAGE RFID-SOLUTIONS, http://www.adage.se/Adage-RFID-Solutions/
13. Kim, T.H., Lee, S.J.: A hybrid hyper tag anti-collision algorithm in RFID system, In: 11th International Conference on IEEE Advanced Communication Technology, ICACT 2009, p. 1276 (2009)
14. Hahnel, D., Burgard, W., Fox, D., Fishkin, K., Philipose, M.: Mapping and localization with RFID technology. In: Proceedings of IEEE International Conference on Robotics and Automation, ICRA 2004, p. 1015 (2004)
15. Darcy, P., Stantic, B., Sattar, A.: A fusion of data analysis and non-monotonic reasoning to restore missed RFID readings. In: 5th International Conference on IEEE Intelligent Sensors, Sensor Networks and Information Processing (ISSNIP 2009), p. 313 (2010)
16. Zhang, S., McCullagh, P., Nugent, C., Zheng, H.: A Theoretic Algorithm for Fall and Motionless Detection. In: Proceedings of the 3rd Annual IEEE International Conference on Pervasive Computing Technologies for Healthcare, pp. 1–6 (2009)
17. Zhang, S., McCullagh, P., Nugent, C., Zheng, H.: Activity Monitoring Using a Smart Phone's Accelerometer with Hierarchical Classification. In: Proceedings of the 6th IEEE International Conference on Intelligent Environment 2010, pp. 158–163 (2010)
18. Chang, C.C., Lin, C.J.: LIBSVM: a library for support vector machines (2001)

Flexible Simulation of Ubiquitous Computing Environments

Francisco Campuzano, Teresa Garcia-Valverde, Alberto Garcia-Sola, and Juan A. Botia

Abstract. Ubiquitous computing software must be reliable. As it occurs in conventional software, one of hardest tasks nowadays is testing. Moreover, if testing is focused on context-aware and adaptive services, the task is even harder. In this case, it is not sufficient testing the software in order to find bugs in the code and repair them (i.e. debugging) and, at the same time, checking that procedures and functions responses to specific interesting inputs are correct. It is also needed checking that the responses of services (i.e. the system under test) to changes in the environment are correct. And the environment includes also human users. This paper proposes UbikSim. An ubiquitous computing environments simulator which tries to alleviate the particularities of testing services and applications whose behaviour depends on both physical environment and users.

1 Introduction

Social Simulation [6] concerns the simulation of social phenomena on a computer using any simulation technique. Software agents used in Multi Agent Based Simulation (MABS) [5] have been used quite extensively to do Social Simulation. In this paper we present Ubiksim [16], a social simulator whose main goal is validating ubiquitous computing services. UbikSim has been designed to simulate environment, domotic devices and people interacting with real ubiquitous software. For this purpose, it is based in MABS. MABS allows representing every agent independently. Since each person is simulated by a separate agent, it is easy to define persons with very different behavioural characteristics. For example, it is possible to produce realistic computational models for people living in a determined environment and making sensors to react on their presence. These devices generate information about the environment (i.e. context information) [7], users and changes in both of them.

Francisco Campuzano · Teresa Garcia-Valverde · Alberto Garcia-Sola · Juan A. Botia
University of Murcia, Spain
e-mail: {fjcampuzano,mtgarcia,agarciasola,juanbot}@um.es

P. Novais et al. (Eds.): Ambient Intelligence - Software and Applications, AISC 92, pp. 189–196.
springerlink.com

With this information, services and applications in ubiquitous computing systems are able to adapt to changes by means of context-aware computing. UbikSim also integrates a middleware for managing contextual information. Such middleware is able to generate some response processes to the simulator depending of the contextual information received. Having a model (i.e. the building, sensors and persons) within a simulation, and its integration with the ubiquitous computing software (i.e. the middleware plus ubiquitous services and a software under test (SUT)), the SUT can be tested. A simulation framework must combine the different elements of simulation into one powerful tool. One of the most important elements of a simulation is validation. UbikSim includes an analyzer whose purpose is validating the simulated models. A social simulator must maximize usability and flexibility while minimising working knowledge of the simulator's engine. For that purpose, the simulator allows researchers to connect multiple SUT to be tested, selecting events of interest for notification. The interaction between SUT and simulator should be performed through standard instruction protocol. UbikSim takes one step towards standardization using an architecture based on a standard for simulators. Standardization is the process of developing and agreeing upon technical standards, that establishes uniform engineering or technical specifications, criteria, methods, processes, or practices. A standard based on a reference model would benefit users and researchers alike, and probably industry would trust more in this kind of applications. In section 2 UbikSim's architecture is defined, section 3 describes how MABS enables Social Simulation in UbikSim, section 4 introduces an example of use. Some related works are described in section 5, conclusions and future works are analyzed in section 6.

2 Architecture

UbikSim's architecture is structured in components. Its basic arrangement is based on a IEEE 1516 High Level Architecture (HLA) Standard [4] requirement, a Run-Time Infrastructure (RTI) that provides a set of services that are necessary to support components to coordinate their operations and data exchange during a run-time execution. This interaction among UbikSim components is based on OSGi [1]. Figure 1 shows the interrelations between components. The elements marked with an asterisk have been designed specifically for UbikSim, while the others are external components that may be replaced in a flexible way. For example, the simulator and the adaptor would change if the MABS platform is changed by other.

In OSGi programming, the elements that can be installed in a framework are called bundles. Each bundle may register zero or more services. Services are Java interfaces, once a bundle registers a service with the OSGi framework, other bundles may use its published service. OSGi provides a great flexibility to UbikSim. Multiple SUTs can be connected to the simulator at the same time (one bundle per SUT). In the client-side there are two bundles. The first, an editor based on Java3D for modelling people and their environment (figure 2(a)). The second bundle is a 3D viewer that shows how the simulation evolves while it is running along the time

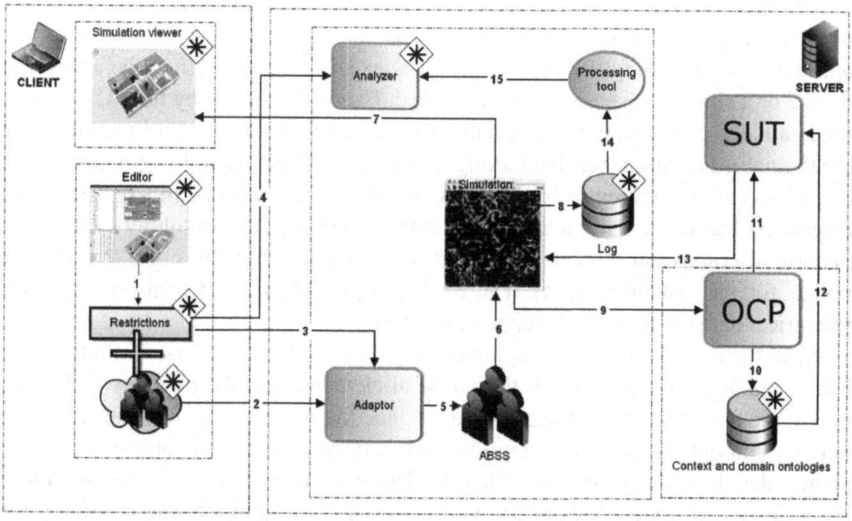

Fig. 1 Collaboration diagram for the UbikSim simulator framework

(figure 2 (b)(c)). In the server-side, the simulation process, the middleware for context management, ubiquitous computing services already validated and the SUT are executed (notice that thanks to OSGi, they can be distributed in different bundles).

The simulator engine of UbikSim is based in a platform that supports MABS. Nowadays, we are using MASON[1], but it is not integral part of UbikSim, it is replaceable by any other MABS platform, for example, Repast[2]. Another feature in UbikSim is a data analyzer whose objective is validating the SUT by means of analyzing the results of the simulations. Validation consists in the determination that the simulated model is an acceptable representation of the real system, for the particular objectives of the model [9].

The second element on the server side is the middleware for context information management. For the current implementation of UbikSim, it is based on Open Context Platform (OCP) [13], developed also in our lab. But others like JCAF[3] may be used. OCP provides support for management of contextual information and merging of information from different sources. It is based on producer-consumer paradigm. The producers introduce information in the system; one or more consumers interested in a particular context information are notified of changes on this information. Context information is stored in an ontology defined according to a context model created by the developer, following the Web Ontology Language (OWL)[4] standard of the W3C.

---

1 `http://cs.gmu.edu/~eclab/projects/mason/`, last access: December 2010.

2 `http://repast.sourceforge.net/`, last access: December 2010.

3 `http://www.daimi.au.dk/~bardram/jcaf/`, last access: December 2010.

4 `http://www.w3.org/TR/owl-features/`, last access: December 2010.

The third element of the server side in UbikSim is compound by the SUT and ubiquitous services already validated. These services are supported to work fine and they are needed by the SUT to properly execute. The SUT may be a physical device or a simulated service to be validated. Events generated by the SUT modify the simulated environment and possibly influence other services, in function of the contextual information received along a simulation. For example, opening a door if a person allowed to enter is near it. A new SUT can be tested in UbikSim if the researcher has previously defined his domain ontology. The ontology contains an historic of contextual information (e.g. it may contain the values of a temperature sensor for the last three days). The SUT can get information from the ontology (historical data) or from OCP (up to date data).

Next, the overall process is explained by following figure 1. First of all, the Ubik Editor bundle is needed to edit the physical elements and the people which compound the environment and agent model. Once the world is created, the editor generates a model specification (step 1) which is passed to the simulator (step 2) with some rules defining restrictions (step 3). These restrictions define what situations are going to generate a SUT's response along the simulation. In the simulator, the first step is adapting the model to the MABS platform underlying UbikSim (step 5). In this case, we have used MASON. Once the Agent Based Social Simulation (ABSS) is defined, the simulation begins to run (step 6). As result of the simulation, we obtain a log history (step 8). UbikSim is OCP'S producer. So, the objects contained in the ontology generates events that OCP manages (step 9) and immediately OCP updates the changes in the domain ontology (step 10). The SUT is connected to OCP and it consumes all contextual information updates in real time (step 11). If the SUT needs some historical information, it queries the ontology database (step 12). When a determined restriction is violated, the SUT sends an standard event to UbikSim that modifies the simulation's environment (step 13). Every modification in UbikSim is updated in the client 3D visor in real time (step 7). An example of this process is introduced in section 4.

After the simulation, the log history can be analyzed. Tools like WEKA[5] may be used to processing the data stored in the log (step 14). The processed data (step 15) and the initial restrictions (step 4) are used as the input for the analyzer. Its function is validating if the SUT has changed properly the simulator's state every time that a restriction have been violated during the simulation.

3 Social Simulation in UbikSim

UbikSim is a bundle which accepts some adaptation commands from a SUT and generates contextual information and a log history as results. It is possible to simulate domotic devices and people in purely social contexts. The behaviours of simulated people are modelled probabilistically. In this approach, behaviours are defined as situations the agent should play in each moment. Transitions between behaviours are probabilistic. The underlying model is a hierarchical automaton (i.e. in a higher

[5] `http://www.cs.waikato.ac.nz/ml/weka/`, last access: December 2010.

level there is a number of complex behaviours that the agent may play and once it is in a concrete state, within the state there is another automaton with more simple behaviours). So, in the lowest level (basic actions), each state is atomic. An agent never conducts two behaviours of the same level simultaneously. Notice that, when the agent is at any state, the necessity of changing to another state may arise. But this is not done immediately. Moreover, a number of different changes (i.e. transitions) may be pending simultaneously. Thus, a list of pending tasks as generated by incoming events is maintained. Such tasks are ordered by a priority. For generating transitions in real time, probability distribution functions are used according to the type of the behaviour and its features. These parametric distributions are known and contrasted with empirical data [8] for concrete environments we already faced. Engineering realistic human models is a hard task. Notice that the level of difficulty is directly related with the level of reality exposed by such kind of models. As the moment, realistic models in UbikSim have been proposed in terms of what the user should be doing at each moment. Thus, the probabilistic automaton mentioned above reproduce such low level of reality. However, a model of what mental process (i.e. an intentional model) should be performing the user now is more interesting and generic. Such kind of models are now being studied.

About domotic devices, every type of sensors or actuators can be modelled. UbikSim offers basic methods to developers for implementing new sensor's configurations. For example, what kind of events they must detect or their range of coverage. Along the simulation, the actuators are going to receive events (i.e. commands) from the SUT. The events are standardized and they allow to modify the most representative attributes of an object (position, color, status...). After a simulation, it is possible to validate if a SUT is working correctly using the analyzer. UbikSim has also been proven validating human behaviours with activity data from real users [8]. If the comparison between real data and produced data is not successfully, the model would be redefined with new parameters to achieve a more realistic model.

4 Use Case

In this section, an example of how UbikSim is being used in a research project called CARDINEA is introduced. The project investigates using ubiquitous computing technologies to augment the quality and safety of work for personnel at hospitals. As specific case is that of hospitals which work against cancer by producing medicaments which imply exposure to low intensity radioactive component during work. A rigorous time of exposure control must be accounted from the hospital's workers. In CARDINEA, it is investigated how RFID and simple services controlling such exposition by person may help in such situations. The hospital's environment is created with Ubik Editor (fig. 2 (a)). Also a restriction is defined, a hospital worker must not be exposed more than the 25% of its daily working turn to radiactive components. The SUT is a service that counts user's exposition time and notifies the system when the restriction is violated.

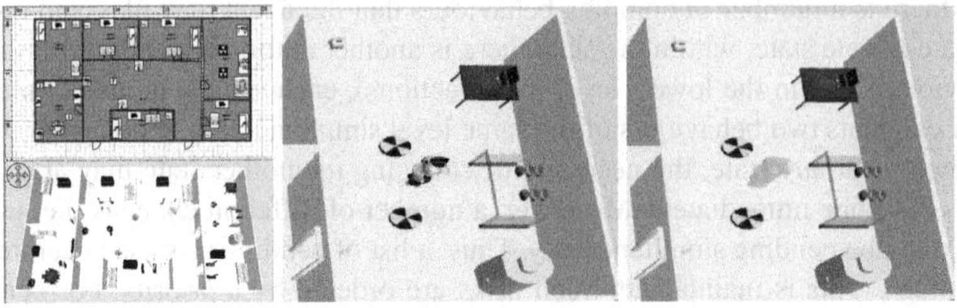

(a) (b) (c)

Fig. 2 (a) A hospital floor created by Ubik Editor. (b,c) A nurse changing along the simulation.

The UbikSim's adaptor creates an agent model comprehensible by MASON and then, the simulator is ready to run. When the simulation is launched, all modifications are shown in the client's 3D viewer in real time. In figure 2 (b) a nurse is inside a room with radioactive elements. A RFID antenna is going to generate contextual information about nurse's position. The SUT receives that contextual information via OCP, and it generates an event that modifies the worker's colour if the restriction regarding exposure to radioactive elements is violated. UbikSim receives this event and it turns the nurse of a green colour in the 3D viewer(fig. 2 (c)). The analyzer may be used for validating if the SUT is working properly. It is possible to reproduce some behaviours which violate the time restriction. The events produced in these reproductions are going to be validated. The analyzer goal is validating if the SUT's notifications are produced when they are necessary.

5 Related Works

Nowadays, various approaches working with simulation within ubiquitous computing application development can be found. One of the most known tool is Ubiwise [2], where real persons generate information about their context, captured through simulated sensors on a simulation engine. TATUS[14] supports simultaneous connection to multiple SUTs using a client-server architecture. There are other works that have followed multi-agent based approaches for defining environments although they do not cover complete ubiquitous computing systems. In Reynolds et al. [15] sensors, actuators, and the environment are simulated, but without simulation of context or adaptable services facilities. Liu et al. [10] uses a distributed intelligent multi-agent system for modelling complex and dynamic user behaviour. In Martin et al. [12] the simulation separates agents, environment and context. UbikSim has the goal of being more complete than the mentioned approaches because it covers environment, context-aware, users and adaptation models. The use of social simulation and MABS provide the possibility of easily integrating user behaviours or relations between them as additional elements to simulate.

In terms of standardization, there are some organizations dedicated to develop simulation standards, for example, Simulation Interoperability Standards Organization (SISO)[6] is an international organization dedicated to the promotion of Modelling and Simulation (M&S) interoperability. Acquisition Operating Framework (AOF)[7] offers a list with all M&S interoperability standards already in use, e.g. HLA. There are two ways choosing a standard to represent an environment model. SEDRIS [11] provides the means to represent any environmental data. In the other hand, the use of ontologies, for example, Standard Ontology for Ubiquitous and Pervasive Applications (SOUPA) [3] is more focused in defining ubiquitous computing environments. All these standards may help UbikSim become a very powerful tool supported by standardization, which gives it an engineering base that distinguishes it when competing with other simulators.

6 Conclusions and Future Works

Ubiksim's architecture satisfies all the requirements needed to offer a flexible simulator. It allows researchers to connect multiple SUT connections to the simulator through OSGi in its client-server architecture. Also it is possible for them to define their own ontology in OCP flexibly and create an environment for their SUT. UbikSim offers basic methods for changing the simulator's environment, and it is possible to select simulator events of interest for notification to the SUT. UbikSim provides a great usability, the editor allows to create maps and to configure objects in it in a very simple way. Running a test is also very easy, UbikSim has a menu where the user can configure parameters like simulation speed. The simulation's viewer is very intuitive, it reflects the current state of domotic devices or people with very intuitive labels or different colours.

Following the IEEE 1516 standard, a HLA-conformant simulator using OSGi has been successfully implemented, but also it is necessary to detail all the possible information which could be sent among components in an Object model template (OMT). This document would provide a common framework for the communication between HLA components. Our target is achieving that all objects would have the same representation both in the editor, UbikSim, and OCP. It would be the next step towards UbikSim standardization. An interesting future work is achieving a standard OCP ontology, and the direct correspondence between objects in the editor, simulator and OCP previously mentioned. An alternative is representing the objects within a complex standard like SEDRIS, which would allow us to define any type of element in the world.

Acknowledgments. This research work is supported by the Spanish Ministry of Science and Innovation in the scope of the Research Projects CARONTE (TSI-020302-2010-129), and DIA++ (TRA2009-0141), through the Fundación Séneca within the Program "Generación del Conocimiento Científico de Excelencia" (04552/GERM/06).

[6] `http://www.sisostds.org/`, last access: December 2010.

[7] `http://www.aof.mod.uk/aofcontent/tactical/mands/content/mands_interoperability_stds.htm`

References

1. Alliance, O.: Osgi service platform, release 3. IOS Press, Inc., Amsterdam (2003)
2. Barton, J., Vijayaraghavan, V.: Ubiwise: A ubiquitous wireless infrastructure simulation environment (2002)
3. Chen, H., Perich, F., Finin, T., Joshi, A.: SOUPA: Standard ontology for ubiquitous and pervasive applications. In: The First Annual International Conference on Mobile and Ubiquitous Systems: Networking and Services, MOBIQUITOUS 2004, pp. 258–267. IEEE, Los Alamitos (2004)
4. Dahmann, J., Morse, K.: High level architecture for simulation: An update. dis-rt p. 32 (1998)
5. Davidsson, P.: Multi agent based simulation: beyond social simulation. Multi-Agent-Based Simulation, 141–155 (2001)
6. Davidsson, P.: Agent based social simulation: A computer science view. Journal of Artificial Societies and Social Simulation 5(1), 7 (2002)
7. Dey, A.K.: Understanding and using context. Personal and Ubiquitous Computing 5, 4–7 (2001)
8. Garcia-Valverde, T., Campuzano, F., Serrano, E., Botia, J.: Human behaviours simulation in ubiquitous computing environments. In: Workshop on Multi-Agent Systems and Simulation at MALLOW 2010 (2010)
9. Law, A., Kelton, W., Kelton, W.: Simulation modeling and analysis. McGraw-Hill, New York (1991)
10. Liu, Y., OGrady, M., OHare, G.: Scalable context simulation for mobile applications. In: Meersman, R., Tari, Z., Herrero, P. (eds.) OTM 2006 Workshops. LNCS, vol. 4278, pp. 1391–1400. Springer, Heidelberg (2006)
11. Mamaghani, F., Foley, P.: SEDRIS-A Collaborative International Infrastructure Technology. In: Proc. Simulation Interoperability Workshop, Paper 04E-SIW-067 (2004)
12. Martin, M., Nurmi, P.: A generic large scale simulator for ubiquitous computing. In: Third Annual International Conference on Mobile and Ubiquitous Systems: Networking & Services, pp. 1–3. IEEE, Los Alamitos (2007)
13. Nieto, I., Botía, J., Gómez-Skarmeta, A.: Information and hybrid architecture model of the OCP contextual information management system. Journal of Universal Computer Science 12(3), 357–366 (2006)
14. O'Neill, E., Klepal, M., Lewis, D., O'Donnell, T., O'Sullivan, D., Pesch, D.: A testbed for evaluating human interaction with ubiquitous computing environments. In: First International Conference on Testbeds and Research Infrastructures for the Development of Networks and Communities, pp. 60–69. IEEE, Los Alamitos (2005)
15. Reynolds, V., Cahill, V., Senart, A.: Requirements for an ubiquitous computing simulation and emulation environment. In: Proceedings of the First International Conference on Integrated Internet ad Hoc and Sensor Networks, p. 1. ACM, New York (2006)
16. Serrano, E., Botia, J., Cadenas, J.: Ubik: a multi-agent based simulator for ubiquitous computing applications. Journal of Physical Agents 3(2), 39 (2009)

Safety Considerations in the Development of Intelligent Environments

Juan Carlos Augusto and Paul J. McCullagh

Abstract. Computing systems are now silently taking part in many of our daily life activities. Intelligent Environments refer to systems which are designed to support humans in daily life. Like any technology these one can fail too. This paper examines potential negative consequences of such systems if they are too naively developed and used and encourages developers in this area to take safety issues into consideration.

Keywords: Intelligent Environments, AAL, reliability, safety critical.

1 Introduction

Building computing systems that operate safely in the real world is very difficult. Compounded by commercial pressure unreliable systems are sometimes expedited and deployed in the marketplace. Even with the best intentions and state of the art resources (both technical and human) it is almost unavoidable that systems contain weaknesses that will lead to failure; it is not a matter of "if" but "when".

Intelligent Environment (IE) systems [Nakashima, 09] are inherently complex, because of the different mix of components (hardware, software, human) that typically compose them. By their very definition, "*...digital environments that proactively, but sensibly, support people in their daily lives.*" [Augusto 07a], these systems are conceived to be deployed in the real world to support humans in a variety of supervisory contexts. Some examples of such systems are "smart" homes, classrooms, cars, offices, manufacturing plants, etc. In some of those applications the artificial system is given an enormous responsibility (e.g., related to safety or well-being).

The magnitude of practical problems to be solved has often concentrated designer's efforts on what to do to get these systems working. Little or no attention is paid to what happens when systems do not behave as anticipated. No company wants to announce that their system at some point will fail to deliver as expected, but it is unavoidable that will eventually happen. Power cuts occur, sensors

Juan Carlos Augusto · Paul J. McCullagh
University of Ulster, United Kingdom
e-mail: {jc.augusto,pj.mccullagh}@ulster.ac.uk

P. Novais et al. (Eds.): Ambient Intelligence - Software and Applications, AISC 92, pp. 197–204.
springerlink.com

malfunction sometimes, sensors can be displaced and the quality of the input to the software taking decisions is degraded. Software can contain bugs, software and hardware updates can introduce errors, or rare, unanticipated and potentially unsafe scenarios can occur. Interoperability issues in complex computing and communication environments can lead to unintended consequences.

As humans start to experience and benefit from the first successful Ambient Intelligent (AmI) systems, in support of their daily activities, the human (traditional) circle of care (family, friends, healthcare professionals) relaxes, invests trust in the system and may not be there when the artificial system fails. Consider the role of AmI in smart homes [Friedewald, 05], in particular providing care for vulnerable people [Orpwood et al, 05]. Evidence from Gaikwad and Warren's systematic review [Gaikwad, 09] demonstrates that home based interventions applied to chronic disease management improve functional and cognitive patient outcomes and reduce healthcare spending. Local authorities work with partners in housing, health, voluntary and independent sectors, and with service users and carers, to implement a telecare-based approach. However, technology-based intervention should not be seen as a substitute for meaningful human interactions and interventions, but as a means of enhancing them.

Everyone involved in the development of IEs have a responsibility to start the discussion on how to design holistically safer systems. Systems (in the broader sense, i.e., the combination of hardware, software, humans and procedures being introduced) should have a responsible attitude towards the environment they serve when they cannot deliver appropriately, and disclose such information in a timely fashion. The AmI community cannot have the concept of an 'accident waiting to happen'. A thread of discussion should be opened within our community on the different ways this can be achieved. Home healthcare is not without risk. Roback [Roback, 03] considered risks that are encountered when placing electronic equipment in this environment. They found that risks and adverse events could stem from technology in itself, from human-technology interaction conditions or from the environment in which the technology is placed. Guidelines were aimed at performance improvement and thus to be considered a complement to the more general guidelines on tele-homecare adopted by the American Telemedicine Association (ATA). Concerns on the safe development and deployment of these technologies were also clearly raised in [Roberts 06].

Thus a major new question arises: Will AmI systems make this form of care safer or potentially dangerous? Next (Section 2) we explain why the state of the art in developing Intelligent Environments should be of concern to all of us and then (Section 3) we propose some additional processes which can be put in place to increase the reliability on these systems.

2 The Argument

This section explains why systems can and will most probably fail at some point and exposes the potential negative consequences for the people these systems are supposed to help.

1. "Hypothesis 1: *Computing Systems DO fail!.*" The history of Software Engineering is plagued with many examples of catastrophic failure (not to mention all the minor faults experience by all of us on a weekly basis by faulty or poorly developed software).

2. "Hypothesis 2: *The more complex the system the more prone to failure.*" Modern computer systems are built as a complex interconnection of specialized modules (Figure 1). As systems become more complex, the potential for failure increases [Holzmann, 03]. The impact of complexity in realiability can be recognized in all fields. For example, Richard Cook [Cook] cites 18 reasons why complexity in the medical system can lead to failure.

3. "Hypothesis 3: *Intelligent Environment systems are inherently complex.*" Intelligent Environments can be developed in any environment where technology can be deployed to assist humans. That infrastructure is supported by a so-called Ambient Intelligence (AmI) that relates software specifically designed to make decisions based on a sensed reality to the technical infrastructure. This creates a complex interpendence and a reliance on several well established areas: sensors, actuators, a (wired or wireless) network, ubiquitous services, HCI, adaptation to users, and accurate reasoning algorithms. Each of these areas are exposed to technical SW or HW faults.

4. "Hypothesis 4: *AmI systems support people. Some of this support is critical (there is a potential for human harm or life loss if the AmI system fails).*" We can potentially consider a wide range of intelligent environments. Some of those that has been started to be explored are: Smart Homes, Smart Classrooms, Smart Cars, health-related applications in hospitals, public transportation in cities, emergency services, industry, decision-making support for business and public surveillance. Smart Homes, for example, are being primarily targeted as a way to increase safety and independence to citizens, in particular of elderly and other vulnerable groups. Failing to detect an unsafe situation or to deliver a call for help through an appropriate channel at the right time can be critical for the person being cared by the system.

5. "Hypothesis 5: *Current state of the art on developing AmI systems is not well organized. In particular, it does not contemplate as a standard that the system may fail.*" The market offers in this area focus on what the system can do and not so much on its limitations and never in its pitfalls. It is not good advertisement for a company to highlight the potential faults systems may have. Still companies should face this topic unashamedly and show genuine interest on offering good and reliable service.

6. "Hypothesis 6: *As humans start to experience the first successful AmI systems supporting their lives the human caring circle relaxes and is not there when the artificial system fails.*" Current caring systems are mostly human based

and rely on professionals from the health system, relatives and friends to care for another human being (Figure 1-a). Imagine the scenario where an Ambient Assisted Living system is deployed and it works acceptably well most of the time to the point that the human carers accept the system and get used to it. As this happens they will feel confident enough to be absent more and more often (Figure 1-b), in some extreme cases this may withdraw completely (Figure 1-c). Still, for example, people with dementia will continue deterioration.

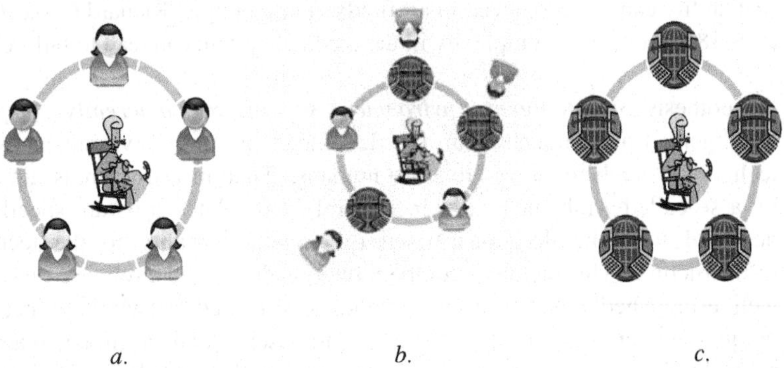

Fig. 1 Potential deterioration of human circle of care

3 Proposed Solutions

The obvious and easy thing to say in these circumstances is: *"Systems should not be built to operate alone"* a sort of "bury your head in the sand" strategy. The problem is people do not necessarily use systems in the ideal form. If a system has been designed to monitor whether an elderly person may have fallen and the system does not work properly, failing to detect or alerting to such an occurrence then regardless of the fact that other carers may or not be available is irrelevant and does not exculpate the responsibility of the system; it is still failing to detect or achieve its main objective. To borrow an example from a different area, McLaren recently recalled baby push chairs in the US as many children had their fingers injured in the folding mechanism. They could have claimed that it was not an intended use or they could have put a warning sticker in an attempt to absolve themselves of responsibility but this solution would not have ameliorated the problem, reduced litigation or have built a credible public reputation. For the same reasons big car manufacturers (Toyota, Honda, and Renault) have recently recalled cars because of the suspicion of faulty mechanisms and as control systems have become more complex, it can be software that contributes to the behaviour of the car.

There is a need for the community developing Intelligent Environments to adopt a more a mature approach to the problem than simply passing the responsibility of problems to the final user. Below there are some suggestions which we recognize are not a definitive word on this subject but may be helpful to initiate a much needed discussion on this topic:

1. *A Formal Software Engineering Approach to AmI Systems Design:* Computer Science, and more specifically its subarea of Software Engineering, have studied since the early days of the discipline the use of systematic methods to increase reliability of software, typically "Testing" and verification by Formal Methods (see [Somerville, 08, Berard, 01; Holzmann, 03; Abdallah, 05]. Traditional Software Engineering Methods can be applied to verify correctness of Intelligent Environments software, see [Augusto 07b, Augusto 09a] for some work in this direction.

2. *The Need for Enhanced understanding of Human Computer (AmI) Interaction:* Further ethnographic research is required to understand how people interact with these systems and use the information they provide, particularly with regard to safety issues. This is required for the person being monitored, the carer and the healthcare professional. Data may be collected through participant observation, interviews, and questionnaires. Human computer interaction experts can contribute to the knowledge base. For previous work that has highlighted the need for methods of validation by users that combine scientific objectiveness with the need of allowing the subjective opinion of the final user of AmI systems, see [Augusto 09b].

3. *A Partnership Between AmI and Human:* Hardware and software should have monitors and reporting built-in. A system of triage may be appropriate. For the most serious errors, the system should conclude that it could cause more harm than good and remove the appearance of a *safety net*[1]. However, for minor errors, it may be possible for the system to work safely with reduced sensors or a faulty software process and continue to work with a reduced capacity. This should be clearly signaled to the users (cared for person, carer and healthcare professional). Where a clean bill of health is given, the system must still monitor the occurrence of unanticipated event that could jeopardize safety.

4. *Hardware:* reliability of the infrastructure sustaining the system is an important dimension in this problem, this encompasses the network as well as the collection of sensing and actuating devices. This infrastructure usually will work reasonably well but networks may cease to work and sensors my malfunction by running out of batteries, being affected by environmental conditions or being moved by the house occupants. Robust methods for the equipment to self-diagnose correct functioning have to be developed.

---

[1] We understand by this the infrastructure and procedures put in place to ensure the fundamental rights and needs of the human looked after by the system are provided.

5. *The Ethical Dimension:* The British Computer Society (BCS) have drawn up seven general principles of ethics in informatics, which we believe should be tested in any AmI system. These fundamental principles, are evaluated in Table 1, and have added relevance where data is collected and processed by complex algorithms; on vulnerable people, some of whom may be dependent on carers or relatives. In particular, AmI systems should also be accountable for any infringement of the privacy rights of the individual person.

Table 1 Informatics ethics and AmI

Ethical Principles	**Factors relating to AmI Systems**
Information Privacy and Disposition.	The cared for person should have control over the information that is collected and made available to carers and relatives.
Openness	The AmI system must be open (so that its decisions can be evaluated), and the information must be made available to the cared for person
Security	Data must be kept securely, particularly as it may be stored for trend analysis and communicated to remote locations.
Access	The data collected should be considered no different to other information in the electronic health record.
Legitimate Infringement	Competing rights of other persons must be respected.
Least Intrusive Alternative	The least intrusive principle applies in AmI systems.
Accountability	AmI system must be open and accountable for any infringement of the privacy rights of the individual person, e.g. alerting a call centre to wandering behaviour

4 Conclusions

AmI researchers have a responsibility to design safer systems, with a high level of transparency. This includes systems that have a responsible attitude towards the environment they serve when they cannot deliver appropriately. Formal specification may provide a means of designing many unsafe conditions out of software. This is time consuming and expensive for normal software, but is important for safety critical AmI applications, and should not be discounted.

It is evident that the hardware and software should be reliable in an AmI system. Thus monitoring is a requirement that is self-testing, periodic checking processes with self report of possible under-performance, e.g. due to a faulty, misplaced sensors or sensors whose battery may need replacing. It may be possible to provide a system which can continue to reason under uncertainty, but this condition must be identified so that periodic human triage can attend to maintenance issues.

We should strive to ensure that AmI used within an assistive environment should improve quality of life. Hence AmI can not only detect alarms, but can become proactive, for example to anticipate abnormal situations and provide guidance for a person under its care. For example the AmI system could be used to guide a person with dementia back to bed during the nighttime [Carswell, 09], if inappropriate or frequent wandering was detected by location sensors. Context, of course, is important to ensure a proper and sensible decision is made. However built into this service model, should always be a human back up. If the AmI systems fails to achieve its objective, then a human carer or friend can be alerted (e.g. via a call centre), and appropriate care restored. This means we should not design a system equivalent to Figure 1 (c), where the human 'safety net' is eliminated, even by stealth or over-confidence in the system.

The sensitivity and specificity of the AmI system then is a key quality metric. If the number of alerts to the users is reduced then the AmI system will add value. However, they should not be reduced to a point where external human help is needed and not signaled by the system, or beyond which the humans become disengaged.

As the capacity of AmI systems increases, and they become interconnected then Web 2.0 technologies (and beyond) may provide human contact with virtual neighbours, and contact with other AmI systems, to rebuild a community feeling and thus enhance the safety net. This of course raises many societal questions with an ethical dimension. What information should be shared and with whom, and will this always benefit the individual being cared for? AmI, like other services, must adhere to the highest ethical principles in support of the human.

Acknowledgments. We acknowledge the valuable contribution of Julie-Ann Augusto from the South Eastern Health and Social Care Trust (UK) made to the content of this paper.

References

1. Nakashima, H., Aghajan, H., Augusto, J.C. (eds.): Handbook on Ambient Intelligence and Smart Environments. Springer, Heidelberg (2009)
2. Augusto, J.C.: Ambient Intelligence: The Confluence of Pervasive Computing and Artificial Intelligence. In: Schuster, A. (ed.) Intelligent Computing Everywhere, pp. 213–234. Springer, Heidelberg (2007)
3. Friedewald, M., Da Costa, O., Punie, Y., Alahuhta, P., Heinonen, S.: Perspectives of ambient intelligence in the home environment. Telematics and Informatics 22(3), 221–238 (2005)
4. Orpwood, R., Gibbs, C., Adlam, T., Faulkner, R., Meegahawatte, D.: The design of smart homes for people with dementia—user-interface aspects. Univ. Access Inf. Soc. 4, 156–164 (2005)
5. Gaikwad, R., Warren, J.: The role of home-based information and communications technology interventions in chronic disease management: a systematic literature review. Health Informatics Journal 15(2), 122–146 (2009)
6. Roback, K., Herzog, A.: Home informatics in healthcare: Assessment guidelines to keep up quality of care and avoid adverse effects. Technology and Health Care 11, 195–206 (2003)
7. Roberts, J.: Pervasive Health Management and Health Management Utilizing Pervasive Technologies: Synergy and Issues. JUCS 12(1), 4–15 (2006)

8. Holzmann, G.: The Spin Model Checker - Primer and Reference Manual. Addison-Wesley Publ., Reading (2003)
9. Cook R. I.: Cognitive technologies Laboratory, University of Chicago. How Complex Systems Fail, `http://www.ctlab.org/documents/How20Complex20Systems20Fail.pdf`
10. Somerville, I.: Software Engineering, 8th edn. Addison-Wesley, Reading (2007)
11. Berard, B., Bidoit, M., Finkel, A., Laroussinie, F., Petit, A., Petrucci, L., Schnoebelen, P., McKenzie, P.: Systems and Software Verification (Model-Checking Techniques and Tools). Springer, Heidelberg
12. Abdallah, A.E., Bowen, J.P., Nissanke, N.: In Dependable Computing Systems: Paradigms, Performance Issues, and Applications, Part I: Models and Paradigms. In: Diab, H., Zomaya, A. (eds.). Series on Parallel and Distributed Computing. J. Wiley & Sons, Chichester (2005)
13. Augusto, J.C., McCullagh, P.: Ambient Intelligence: Concepts and Applications. Int. Journal on Computer Science and Information Systems 4(1), 1–28 (2007)
14. Augusto, J.C.: Increasing Reliability in the Development of Intelligent Environments. In: Proc.of 5th Int. Conf. on Intelligent Environments (IE 2009), Barcelona, pp. 134–141 (2009)
15. Augusto, J.C., Bohlen, M., Cook, D., Flentge, F., Marreiros, G., Ramos, C., Qin, W., Suo, Y.: The Darmstadt Challenge (the Turing Test Revisited). In: Proceedings of the Int. Confer-ence on Agents and Artificial Intelligence (ICAART), Porto, Portugal (2009)
16. Carswell, W., McCullagh, P.J., Augusto, J.C., Martin, S., Mulvenna, M.D., Zheng, H., Wang, H.Y., Wallace, J.G., McSorley, K., Taylor, B., Jeffers, W.P.: A Review of the Role of Assistive Technology for People with Dementia in the Hours of Darkness. Technology and Health Care 17(4), 281–304 (2009)

LECOMP: Low Energy COnsumption Mesh Protocol in WSN

Juan A. Ortega, Alejandro Fernandez-Montes, Daniel Fuentes,
and Luis Gonzalez-Abril

Abstract. LECOMP is a mesh network protocol that is focused on minimizing battery consumption of nodes in a Wireless Sensor Network (WSN). This paper is a improvement of a first approach [1] which proposes a method to extend the lifetime and the reliability in a WSN balancing transmissions saving energy throughout different nodes in order to consume homogeneously nodes' batteries, and taking into account distances and hops of routes. First, a central node analyzes messages from nodes during a training period to determine new routing rules for each sensor. Second, the server configures the network adding the computed rules to the nodes. Third, central server can reconfigure the network routes when new balance of load is needed.

1 Introduction

Since sensors are battery powered in Wireless Sensor Networks applications, it is essential to minimize energy consumption so lifetime of the global network can be extended to reasonable times. The establishment of WSNs have grown in both research and real-world applications so improvements and solutions are required. Nowadays, energy-efficient sensor networks is a common topic in research works and the main directions to energy conservation are discussed. Some studies deal with the topology control which refer to the techniques that are aimed to super-imposing a hierarchy on the network organization to reduce the consumption [2]. In this sense, the location and the distance between the nodes are critical aspects. Some studies determines that multi-hop communication in mesh networks consumes less power than the traditional single-hop long distance radio communication in star networks

Juan A. Ortega · Alejandro Fernandez-Montes · Daniel Fuentes
Languages and Computer Systems Department, University of Seville, Seville, Spain
e-mail: {jortega, afdez, dfuentes}@us.es

Luis Gonzalez-Abril
Applied Economics I Dept, University of Seville, Seville, Spain
e-mail: luisgon@us.es

P. Novais et al. (Eds.): Ambient Intelligence - Software and Applications, AISC 92, pp. 205–212.
springerlink.com

[3]. Other studies propose an efficiency power management in the network using sleep/wakeup protocols at the application or network layers [4, 5], or strictly integrated MAC [6] protocols. There are multiple application environments of energy-efficient wireless sensor networks and different routing techniques for them [7, 8]. Others methods save energy by reducing the number of sensor transmissions. In [9], sensor transmissions are sorted according to the magnitude of their measurements, and the sensors with small magnitude measurements, less than a threshold, do not transmit. However, in [10] the data compression allows the reduction of the data length. On the other hand, battery analysis is another approach to balance the transmissions and maximize lifetime of the network[11]. In this paper some of the described techniques are combined to achieve a energy-efficient communication in a mesh WSN. Specifically, the batteries levels, sensor distances and message routes are analyzed to balance load and to optimize the routes from nodes to the central station. In Section 2 we provide an overview of the objectives then, in Section 3, we describe our energy-efficient routing proposal LECOMP. Implementation and test results are related in Section 4 and finally, concluding remarks are made in Section 5.

2 Routing Proposal

The primary goal of LECOMP protocol for WSN is to increase the lifetime of the network saving batteries. Moreover, secondary goals are: decreasing the load of the netwrok by minimizing configuration messages needed to setup sensors routing constraints in an automated and live way, and on the other hand, to free the administrator from manually setup the network. The architecture of our sensor network is illustrated in figure . We consider next elements:

- A set of n motes or sensors, $S_1, S_2, ..., S_n$. Devices that sense on aspects of the physical world like the humidity or temperature levels. These sensors are typically battery powered and this information can be spread throughout the network.
- The central server (aka end node), H that stores aspects of the mote network state and acts as a proxy to communicate between mote network and client applications.
- A set of client applications, A that interfaces with the host server to communicate with nodes, to manage existing applications or deploy new ones.

We assume that sensors initially interact each other through a standard mesh network that tries to maximize the reliability of the network broadcasting packets to all nodes in order to reach the end node.

A message from a sensor can be captured by all sensors that are into its scope. The server can communicate with sensors within its scope, but client applications only communicate with the server. The size of a message is limited but it can be configured and modified to include or exclude information.

The main goal of this routing proposal is the optimization of the number of messages copies traveling through the network and the improvements of routes in order

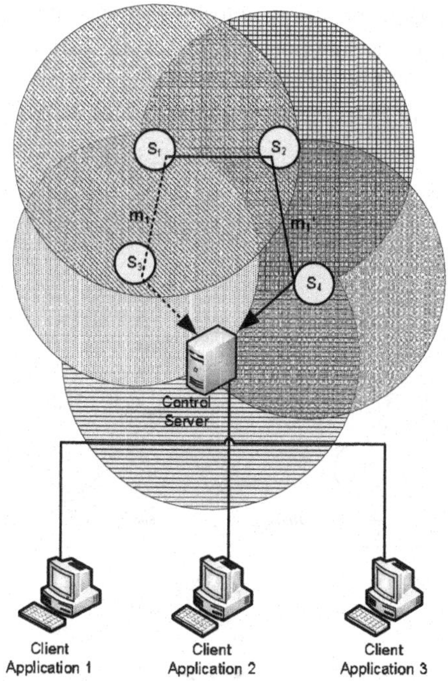

Fig. 1 Representation of typical WSN. A message is sent from S_1 node to H server.

to save batteries in the nodes by a previous training period. This way the H server computes rules for each sensor to determine when is necessary to forward a received message.

When the training-time begins, all the sensors start sending messages to the server. In every message, the description of the route (nodes list, where a message has been routed) is attached to the message. When a message arrives to the H server, the message contains the whole route from the origin sensor to the server H. With this information, the H server can establish some rules to improve the communication between the nodes. This rules are computed by an implementation of the A* algorithm for pathfinding. A* algorithm [12] uses a distance-plus-cost heuristic function $f(x)$ which is a sum of two functions:

- $g(x)$ The path cost function. We assume this function returns the standardized weight (power in watts needed) of the edge between two nodes. This weight is the quantity of power needed to make the transmission between two nodes.
- $h(x)$ The heuristic function. Our solution uses the minus geometric mean of battery levels (in [0-1]) of current path. This way we are taking into account battery levels of a path in order to benefit paths where battery levels of nodes are higher. Notice that geometric mean is used to detect paths where at least one of the nodes has a very low battery level.

Comparing with Dijkstra or others algorithms for pathfinding, A* includes both cost function and heuristic function so it mixes depth-first search and breadth-first search. A* is equals to Dijkstra algorithm if $h(x) = 0$. For example, let's consider two routes R_1 and R_2 of two copies m_1 and $m2$ of the same message respectively:

$$R_1 = [S_1, S_2, S_4, H]$$

$$R_2 = [S_1, S_3, S_4, H]$$

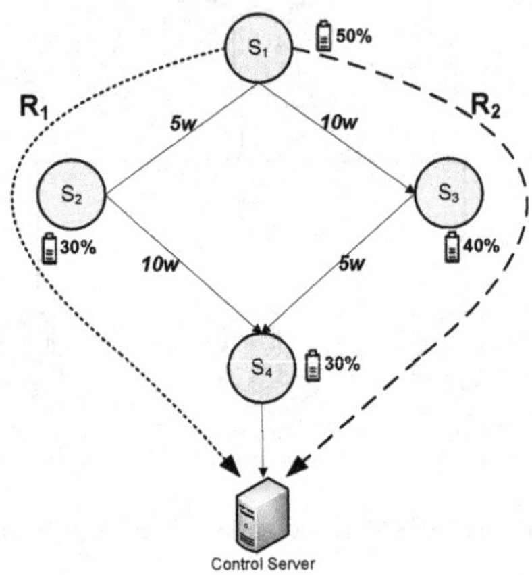

Fig. 2 Case study for 4 nodes and two possible paths

Nodes, routes, costs and battery levels are shown in figure 2. Once the server receives paths, costs and battery levels computes best paths. In the situation shown route R_1 final cost is computed bellow:

$$f(1) = g(1) + h(1) = 0 - 0.5 = -0.5$$
$$f(2) = g(2) + h(2) = 0.5 - \sqrt[2]{0.5 \cdot 0.3} = 0.11$$
$$f(4) = g(4) + h(4) = 1 - \sqrt[3]{0.5 \cdot 0.3 \cdot 0.3} = 0.64$$
$$Cost(R_1) = f(1) + f(2) + f(4) = -0.5 + 0.11 + 0.64 = \mathbf{0.25}$$

And for route R_2:

$$f(1) = g(1) + h(1) = 0 - 0.5 = -0.5$$
$$f(3) = g(3) + h(3) = 1 - \sqrt[2]{0.5 \cdot 0.4} = 0.56$$
$$f(4) = g(4) + h(4) = 0.5 - \sqrt[3]{0.5 \cdot 0.4 \cdot 0.3} = 0.1$$
$$Cost(R_2) = f(1) + f(3) + f(4) = -0.5 + 0.56 + 0.1 = \mathbf{0.16}$$

With these paths, the server can detect that it is not necessary that the S_3 sensor forwards the message m_2 (which is a copy of m_1). Hence, the rule *"Don't forward a message if the previous nodes sequence is* $[S_1]$*"* will be established for the S_3 sensor when the server reconfigure the network. In this manner, the server H continues receiving messages and computing new rules until a time threshold is exceeded. In this case, the nodes stop sending messages to the server and the reconfiguration of the nodes starts. In this stage, the server establishes the computed rules to the nodes and the network starts using this new configuration.

This way we can achieve energy saving by three actions:

- reducing message hops and avoiding copies.
- prioritizing the route where costs are lower.
- balancing load to maximize general battery levels.

Obviously, during the training time, the decrease of the energy consumption is null. However, after this period, the key limitations of WSNs, the storage, power and processing, are treated. The server reconfigures the nodes with the calculated rules and the communication in the network. With this optimization, in the last example the message m_2 will not arrive to server. Hence, duplicated messages in the network are avoided and consequently, the information load in buffers, the time processing and the battery consumption in sensors are reduced and, consequently, the lifetime of the network is increased. Furthermore, by deleting data redundancy and minimizing the number of data transmissions, there is less data load in the network, the possibilities of an overload are lower and the message management is simplified.

```
// Called when the Dispatcher receives a message for this
   protocol.
begin public void stackReceive [Receiver rcvr]
    // Gets the message
    RoutedMsg msg=(RoutedMsg)rcvr.getData();
    // Check hasn't been received already and route is
       allowed
    if (!msg.equals(prevMessage) &&
    !msg.currentRoute.equals(restrictedRoute)) then
        // Adds current sensorId to the route
        msg.currentRoute.addElement(id);
        // Not-relevant operations
        TestReceiver meta=new TestReceiver();
        meta.addMetadata(rcvr);
        meta.data = msg.data;
        // This will forward the received data
        ds.dispatch(meta);
        prevMessage = msg;
    end
end
```

Algorithm 1. Routine running on sensor nodes that checks if message forwarding is needed

As batteries drain the H server computes new rules, and periodically, the H server reconfigures the routing rules to balance batteries of the whole network. In addition to this reconfiguration if a node stops working due to an attack, the lack of battery or another external factor it will not send any reply. In this case, the H server will check affected paths and compute new paths to replace invalid ones.

3 Implementation, Tests and Results

For test purposes we have implemented the routing proposal in Sentilla Work, the IDE supplied with Sentilla Development Kit. The real implementation is an extension of one of the protocols supplied by Sentilla, therefore the algorithm shown it was written in Java.

Results are shown in Table 1 below where the base mesh protocol, the first version of the method called LECOMP-1 [1] and present method LECOMP-2 are compared. Tests have been done at the WSN deployed at Computer and Systems Languages department of the University of Sevilla. It consists on 9 Zigbee sensors deployed in 9 offices. The number of avoided messages in LECOMP-1 and LECOMP-2 is calculated through an approximation between the number the time computed and the duplicated messsages received in the base case. Table 1 shows that in case of LECOMP-1 the sensor network lifetime is 421h and for LECOMP-2 439h. Compared to 360h when no LECOMP is applied this is an improvement of approximately 17% in case LECOMP-1 and 22% of LECOMP-2 in terms of sensor network lifetime. That implies that when LECOMP-1 is used the WSN runs 2,5 days longer and in case of LECOMP-2 over 3 days longer than when no intelligent routing applied. Moreover, we would like to emphasize the improvement of about 7% in avoided messages and 10% in avoided steps in this proposal comparing with the first version of the method. In figure 3, the battery life and the (no-duplicated) received messages

Table 1 Base-Protocol, LECOMP-1 and LECOMP-2 comparison

	Base	LECOMP-1	LECOMP-2
#sensors	9	9	9
timeComputed	360h.	421h.	439h.
#msgsPerHour	30	29.5	29.7
#msgs	10800	12425	13044
trainingTime	0h.	1h.	1h.
#duplicatedmsgs	1944	5	5
avgStepsPerMsg	2,3	2,02	2,12
#msgsReceived	12744	12430	13049
#msgsAvoided	0	2249	2945
%avoidedMsgs	0	**18,1**	**22,5**
#Steps	29311	25524	27933
#StepsAvoided	0	7368	9020
%avoidedSteps	0	**25,1**	**35,9**

in the three cases are compared. Using our method the number of received messages and the lifetime of the WSN increments.

Figure 3 the battery life and the (no-duplicated) received messages in the three cases is compared. Using our method the number of received messages and the lifetime of the WSN increments.

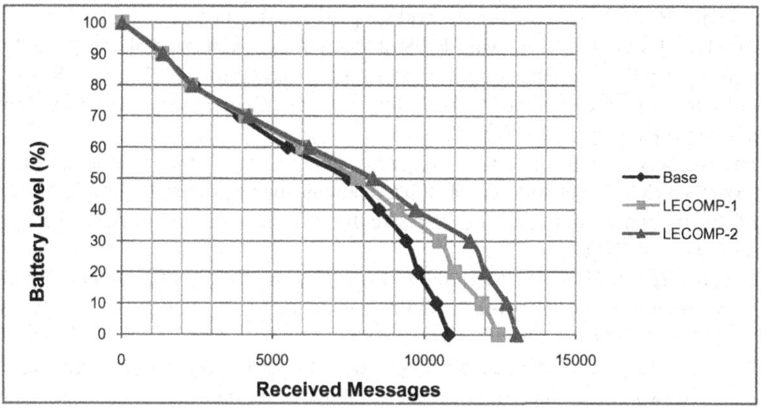

Fig. 3 Comparison Graphic between Base Mesh protocol, LECOMP V1 and LECOMP V2

4 Conclusion

In this paper, we have presented a energy-efficient method for mesh wireless sensor networks. We have got to save up energy by reducing the forwarding of messages. First, the central server determines the rules through the information of the messages routes, the battery levels and the distance between sensors. Second, the network is reconfigured avoiding duplicated messages and balancing the load. The described practical experiences with over Sentilla sensors and Zigbee technology shows an improvement in the WSN lifetime of 22% compared with the base protocol.

Acknowledgements. This research is supported by the MCI I+D project ARTEMISA (TIN2009-14378-C02-01).

References

1. Fuentes, D., Fernandez-Montes, A., Ortega, J.A., Gonzalez-Abril, L.: Energy-Efficient Routing Approach for Wireless Sensor Networks. In: Proceedings of the 4th Symposium of Ubiquitous Computing and Ambient Intelligence UCAmI 2010, pp. 9–12 (2010)
2. Shrestha, A., Xing, L.: A Performance Comparison of Different Topologies for Wireless Sensor Netwoks. In: IEEE Conference on Technologies for Homeland Security 2007, pp. 280–285 (2007)

3. Chandrakasan, A., Min, R., Bhardwaj, M., Cho, S.H., Wang, A.: Power Aware Wireless Microsensor Systems. In: Keynote Paper. 28th European Solid-State Circuits Conference (ESSCIRC), pp. 47–54 (2002)
4. Anastasi, G., Conti, M., Di Francesco, M., Passarella, A.: Energy conservation in wireless sensor networks: A survey. In: Proceedings of the 33rd Hawaii International Conference on System HICSS 2000, pp. 10–20 (2000)
5. Mathur, G., Desnoyers, P., Chukiu, P., Ganesan, D., Shenoy, P.: Ultra-low power data storage for sensor networks. In: Proceedings of the 5th International Conference on Information Processing in Sensor Networks, pp. 374–381 (2005)
6. Shih, E., Cho, S., Ickes, N., Min, R., Sinha, A., Wang, A., Chandrakasan, A.: Physical Layer Driven Protocol and Algorithm Design for EnergyEfficient Wireless Sensor Networks. In: Proceedings of the 7th Annual International Conference on Mobile Computing and Networking, pp. 272–287 (2001)
7. Pantazis, N.A., Nikolidakis, S.A., Vergados, D.D.: Energy-Efficient Routing Protocols in Wireless Sensor Networks for Health Communication Systems. In: Proceedings of the 2nd International Conference on PErvsive Technologies Related to Assistive Environments, vol. 34 (2009)
8. Zeng, K., Ren, K., Lou, W., Moran, P.J.: Energy Aware Efficient Geographic Routing in Lossy Wireless Sensor Networks with Environmental Energy Supply. In: Proceedings of QShine 2006 (2006)
9. Chen, X., Blum, R., Sadler, B.M.: A New Scheme for Energy-efficient Estimation in a Sensor Network. In: Proceedings of the Conference on Information Sciences and Systems, CISS (2009)
10. Marcelloni, F., Vecchio, M.: A simple algorithm for data compression in wireless sensor networks. IEEE Communications Letters 12(6) (2008)
11. Tang, Q., Yang, L., Giannakis, G.B., Qin, T.: Battery Power Efficiency of PPM and FSK in Wireless Sensor Networks. IEEE Trans. on Communications 6(4), 1308–1319 (2007)
12. Hart, P.E., Nilsson, N.J., Raphael, B.: A Formal Basis for the Heuristic Determination of Minimum Cost Paths. IEEE Transactions on Systems Science and Cybernetics 4(2), 100–107 (1968)

A User-Friendly Interface for Rules Composition in Intelligent Environments

Dario Bonino, Fulvio Corno, and Luigi De Russis

Abstract. In the domain of rule-based automation and intelligence most efforts concentrate on building the technological infrastructure, often disregarding user-home interaction requirements. This paper attempts to mitigate this issue by defining a rich-web rule visual design interface specifically aimed at non-skilled home inhabitants.

Keywords: Domotics, Rules Composition, Intelligent Environment, Human-Home Interaction, User Interface, Smart Home, Interface Concept.

1 Introduction

Many intriguing scenarios are currently sketching the home of the future, where human inhabitants will only carry out "exciting" or "interesting" tasks and the home will take care of all boring duties that fill every day life. Although appealing, this long-term vision (part of the Ambient Intelligence research field) has also a worrying connotation where homes not only facilitate our life but directly modify our home-related behavior in ways difficult to forecast, on the user side. This scenario, already emerging from several studies about user attitudes towards smart homes [1, 3], has been driving an initial research effort on finding suitable trade-offs between totally direct user control and fully automatic home behaviors, involving several degrees of home autonomy.

No sound and widely agreed solution to this trade-off has currently been found and the related research activities, both in the Human Computer Interaction (HCI) and Ambient Intelligence (AmI) communities are still very active. Nevertheless, a relatively accepted approach based on *activity delegation* is gaining momentum. In the AmI community, explicit delegation of tasks to homes is usually realized

Dario Bonino · Fulvio Corno · Luigi De Russis
Politecnico di Torino, Dipartimento di Automatica ed Informatica,
Corso Duca degli Abruzzi 24, 10129 - Torino, Italy
e-mail: {dario.bonino, fulvio.corno, luigi.derussis}@polito.it

P. Novais et al. (Eds.): Ambient Intelligence - Software and Applications, AISC 92, pp. 213–217.
springerlink.com

through rule-definition or user-initiated learning. While technology, especially for rule-based delegation, is rather mature and widely investigated, there is still a sensible lack of effective user interfaces. To support users in shaping their specific home automation policies, interfaces must be simple, easy to use and to learn for people without advanced computer skills, and should not require any specific notion about the automation technology installed in the home.

In this paper we propose a first step for overcoming the current lack of effective rule definition interfaces by proposing a paper prototype of a rich-web rule design interface specifically aimed at non-skilled home inhabitants. The remainder of the paper is organized as follows: Section 2 relates the proposed approach to the state of the art, highlighting commonalities and differences. Section 3 defines requirements for rule creation interfaces dedicated to non-skilled home inhabitants, while Section 4 introduces the interface design. Section 5 concludes the paper and depicts future works and research scenarios.

2 Related Works

Different rule composition interfaces, aimed at people without technical skills, are present in the literature, either for domestic environments or for other application domains. An example of such interfaces is OpenBlocks. *OpenBlocks* [4] is a general-purpose framework for graphical block programming systems, developed at the MIT. This framework is more expressive than the interface we propose and could represent a complementary solution to realize a rule builder. However, OpenBlocks requires a heavy customization to be adapted for a domotic environment, it is not web-based and its higher expressiveness leads to increased complexity for rules creation.

Most of rule builder applications specifically developed for domestic environment are *context-aware* applications, such as iCAP [2]. The *iCAP* interface has one window with two main areas. A tabbed window is the repository for user-defined inputs, outputs, and rules. The input and output components are associated with graphical icons that can be dragged into the main area, to be later used to construct a conditional rule statement. An example rule could be: *IF Sam is in the office after 5pm and the temperature is less than 10 Celsius degrees OR IF Jane is in the bedroom and the temperature is between 0 and 15 Celsius degrees, THEN turn on the heater in the house*. The grammar used by iCAP is similar to the one proposed by our interface, but iCAP is not web-based, requires the user to draw each object she wants to use in a rule, is pen-based and does not differentiate between events and conditions.

3 Requirements

Delegating part of everyday tasks to the home requires suitable interfaces for enabling the home inhabitants to easily define processes to be automated, i.e., to effectively program automation rules. By interacting with both people living and managing smart homes and with people commercializing wired and wireless home automation systems we derived the following set of requirements that an effective rule builder interface shall obey.

1. Rules shall be ***definable by people with basic level of computer literacy***, the only required knowledge is about the home components, in terms of normal usage and behavior.

   a. Home *devices shall be exposed in an abstract and technology independent way*, thus enabling user to easily specify the rule objects.
   b. *Rules shall be self-explaining*, i.e., they can be directly/easily translated in a nearly natural language description.
   c. *Rules shall always be "valid"*, i.e., the user can only create and save syntactically (and possibly semantically) correct rules.
   d. Rules shall be *expressive enough to manage most situations, actions and interactions* that a home inhabitant may want to delegate (i.e., they shall be easily mapped onto a powerful enough computer-based rule language).

2. Rules shall be defined by using a ***wide range*** of possible ***interface devices*** (e.g., PCs, touch screen panels), and ***input modalities*** (e.g., touch, mouse).
3. The ***rule-design interface must facilitate*** the ***delegation of tasks*** from humans to homes ***providing suitable "aids"***.

   a. Rules editing shall be facilitated by means of *suggestions, guiding interfaces* and *auto-filling* functionalities.
   b. Rule interface should offer *support to handle unexpected loss of connections* or *computer malfunctions*, e.g., *automatically saving rules*.

4 Design

Interface Concept

Sam is a smart home user with little technological skills. Thanks to a web application, he configured the devices present in his home, renaming them at his pleasure. Now, he wants to create a rule such as "*if the living room is dark, turn on the lamp*". Sam opens his browser to reach the Rule Builder. The interface he sees on his screen is sketched in Figure 1. On the left, he sees what he needs to create the rule, including a lamp and a light sensor, previously added from the house map. On the right, he can see a wide area to be used for the definition of a rule. The dotted rectangles under the IF and THEN keywords are strong visual clues suggesting to drag a device inside them (req. 3a). Sam decides to drag the light intensity sensor under the "IF". By dragging the icon, it docks under the "IF" as a rectangular container. In this container, besides the sensor name, Sam sees a list to specify what sensor event should be intercepted. He chooses "LOW" light intensity (req. 1a).

Sam, after, drags the lamp icon under the "THEN" keyword. When he starts to drag the icon, two other rectangles appear before the "THEN" keyword (req. 3a and

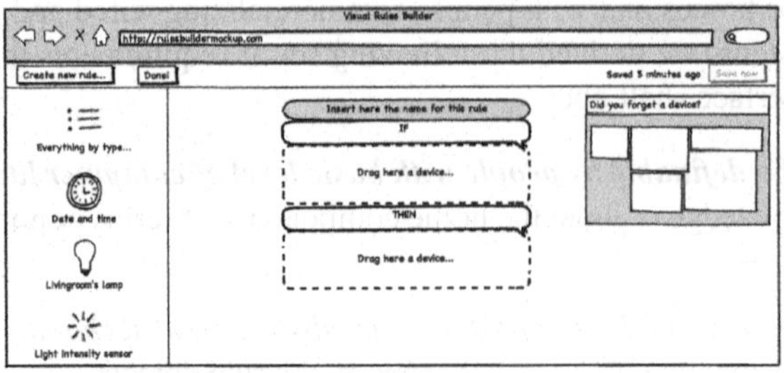

Fig. 1 The main page of our Rule Builder

1c): the optional "WHEN" and "OR IF" statements (req. 1b and 1d). He continues to drag the lamp icon under the "THEN" keyword and selects "ON" between the options presented by the lamp container (req. 1a).

The rule is complete (Figure 2) and the lower part of the interface reports a sentence that summarizes the just created rule (req. 1b).

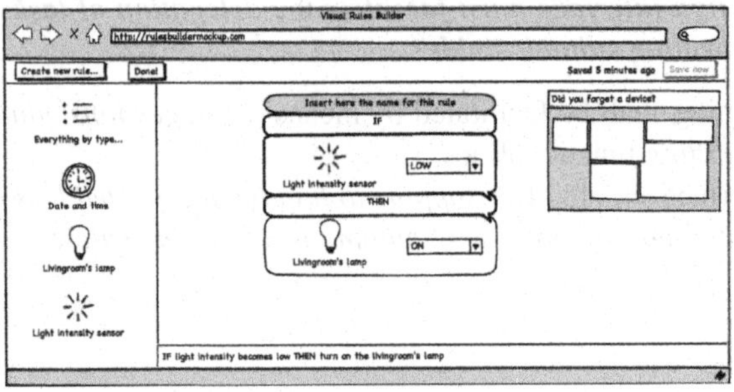

Fig. 2 Complete rule

Grammar

The Rules Builder concept just illustrated guarantees rule correctness (req. 1c) and readability (reqs. 1b, 1d and 3a) by exploiting a formal rule representation grammar (see Figure 3) based upon four fixed keywords: IF, THEN, WHEN, OR IF (req. 1c and 1d). The first two are mandatory for the creation of any rule, while the others are optional (dotted in Figure 3). A rule composed with this grammar follows the natural language (req. 1b). The IF keyword expresses an event to trigger the rule. The event is indicated, in Figure 3, as an "E-BLOCK" (event-block). WHEN defines one or more conditions constraining the event; multiple constraints should be simultaneously satisfied. The set of constraints is shown as "C-BLOCKS" (constraint-blocks). OR IF is a disjunction for repeating the IF-WHEN part more than once. Finally, THEN indicates a set of actions to be executed on the occurrence of the above triggers. The actions are indicated as "A-BLOCKS" (action-blocks).

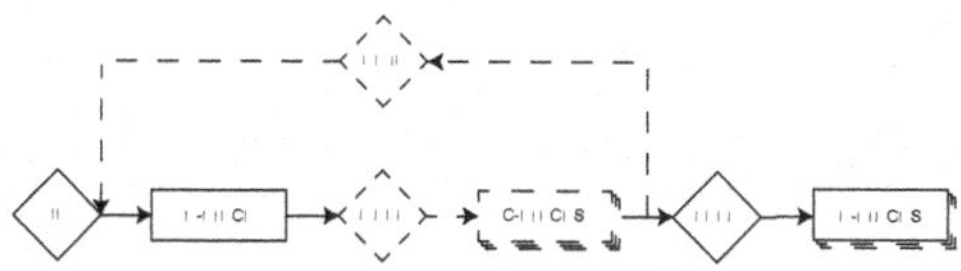

Fig. 3 The grammar underlying the creation of a rule

To maintain rule consistency (req. 1c), each device involved in the creation of a rule has a different behavior according to the block in which is inserted: E-BLOCK interprets events generated by controllable devices, clock and sensors; C-BLOCK supports controllable devices, clock and sensors; A-BLOCK, finally, supports controllable devices only.

The interface concept and the Rules Builder grammar have been informally verified, and approved, by a restricted test group with basic computer skills.

5 Conclusions and Future Works

This paper distills the basic requirements of home rule development environments and introduces the conceptual design of Rule Builder, a rich-web interface that specifically targets non-expert users, i.e., home inhabitants with little or no technological skills. Future works are the realization of a working prototype fulfilling all the requirements proposed and its integration with Dog, an ontology-based domotic gateway[1]. Extensive experimentation with users will then provide means for a sound validation of the proposed interface.

References

1. Demiris, G., Rantz, M., Aud, M., Marek, K., Tyrer, H., Skubic, M., Hussam, A.: Older adults' attitudes towards and perceptions of "smart home" technologies: a pilot study. Medical Informatics and the Internet in Medicine 29(2), 87–94 (2004)
2. Dey, A., Sohn, T., Streng, S., Kodama, J.: iCAP: Interactive prototyping of context-aware applications. In: Fishkin, K.P., Schiele, B., Nixon, P., Quigley, A. (eds.) PERVASIVE 2006. LNCS, vol. 3968, pp. 254–271. Springer, Heidelberg (2006)
3. Eggen, B., Hollemans, G., van de Sluis, R.: Exploring and Enhancing the Home Experience. Cognition, Technology and Work 5(1), 44–54 (2003)
4. Roque, R.V.: OpenBlocks : an extendable framework for graphical block programming systems. Massachusetts Institute of Technology (2007), `http://hdl.handle.net/1721.1/41550`

[1] `http://domoticdog.sourceforge.net`

Using an Ambient Intelligent Architecture for Developing an Intelligent Tutoring System for Training Operators of Power System Control Centres

Luiz Faria, António Silva, Carlos Ramos, Luís Gomes, Zita Vale, and Albino Marques

Abstract. This paper describes an architecture conceived to integrate Power Systems tools in a Power System Control Centre, based on an Ambient Intelligent (AmI) paradigm. This architecture is an instantiation of the generic architecture proposed in [1] for developing systems that interact with AmI environments. This architecture has been proposed as a consequence of a methodology for the inclusion of Artificial Intelligence in AmI environments (ISyRAmI - Intelligent Systems Research for Ambient Intelligence). The architecture presented in the paper will be able to integrate two applications in the control room of a power system transmission network. The first is SPARSE expert system, used to get diagnosis of incidents and to support power restoration. The second application is an Intelligent Tutoring System (ITS) incorporating two training tools. The first tutoring tool is used to train operators to get the diagnosis of incidents. The second one is another tutoring tool used to train operators to perform restoration procedures.

Keywords: Ambient Intelligence, Artificial Intelligence, Context Awareness, Intelligent Tutoring Systems.

1 Introduction

Ambient Intelligence (AmI) deals with a new world where computing devices are spread everywhere, allowing the human being to interact in physical world environments in an intelligent and unobtrusive way. These environments should be aware of the needs of people, customizing requirements and forecasting behaviours. AmI environments may be so diverse, such as homes, offices, meeting

Luiz Faria · António Silva · Carlos Ramos · Luís Gomes · Zita Vale
GECAD – Knowledge Engineering and Decision Support Group
Institute of Engineering – Polytechnic of Porto, Portugal

Albino Marques
REN – Transmission Network, Portugal

P. Novais et al. (Eds.): Ambient Intelligence - Software and Applications, AISC 92, pp. 219–226.
springerlink.com

rooms, schools, hospitals, control centres, transports, touristic attractions, stores, sport installations, music devices, etc. In the aims of Artificial Intelligence (AI), research envisages to include more intelligence in the AmI environments, allowing a better support to the human being and the access to the essential knowledge to make better decisions when interacting with these environments. However, today some systems use AmI like a buzzword, and most of times a limited amount of intelligence is involved. Some researchers are building AmI systems without AI, putting the effort just in operational technologies (sensors, actuators, and communications). However, soon or later, the low level of intelligence will be a clear drawback for these systems. The acceptability of AmI will result from a balanced combination from AI and operational technologies. Figure 1 has been published in [1] and shows a view of Ambient Intelligent Systems.

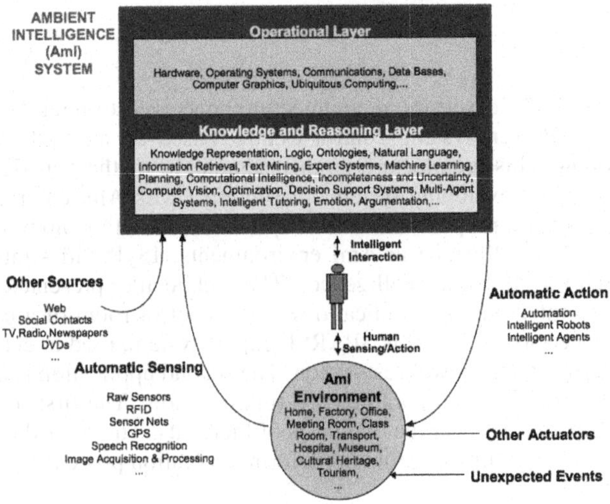

Fig. 1 Ambient Intelligence view from Artificial Intelligence perspective

2 ISyRAmI Methodology

The main objective of ISyRAmI (Intelligent Systems Research for Ambient Intelligence) is to develop a methodology and architecture for the development of AmI systems considering AI methods and techniques, and integrating them with the operational part of the systems. It is important to notice that the ISyRAmI architecture is centred in the concepts and is oriented for the intelligent behaviour, in spite of being oriented for technologies. In this way, it makes no sense to compare ISyRAmI with other technology-oriented approaches previously proposed for AmI.

AI methods and techniques can be used for different purposes in AmI environments and scenarios. Several AI technologies are important to achieve an intelligent behaviour in AmI systems. Areas like Machine Learning, Computational

Intelligence, Planning, Natural Language, Knowledge Representation, Computer Vision, Intelligent Robotics, Incomplete and Uncertain Reasoning, Multi-Agent Systems, and Affective Computing are important in the AmI challenge [2].

In the AmI view illustrated by Figure 1 the data/information/knowledge for the AmI system arrives not only by the automatic sensing, but also from other sources (web, social contacts, media, etc). Besides, the automatic action is not directed just to the AmI environment, but also for the other actuators of the AmI environment (other persons or autonomous devices) and to the other sources (for example for updating an information). The external events are separated in Other Actuators (e.g. other persons or autonomous devices in the same environment) and Unexpected Events (e.g. a storm). The interaction between the human being supported by our AmI system and the other actuators is now clearer.

3 AmI Architecture for Power System Control Centers

Before describing the internal ISyRAmI architecture we will centre our attention in the connection with the external parts. Our AmI system is able to obtain data, information and knowledge from two distinct sources:

- from the AmI environment, using a set of automatic sensing devices (raw sensors, cameras, microphones, GPS, RFID, sensor nets);
- from other sources, here we include the web, social contacts, general persons, experts, radio, TV, newspapers, books, DVD, etc.
- Our AmI system may act in the four different ways:
- operating directly on the AmI environment, using automation devices, robots, and intelligent agents;
- interacting with human beings by means of a decision support system, and the actuation on the AmI environment is done by this/these person(s);
- interacting directly with the other external actuators with the ability to act on the AmI environment (e.g. asking one person or robot in order to do some specific task on the environment);
- updating the other sources of data/information/knowledge.

We also assume that the human beings supported by the AmI system have access to the other sources and can interact themselves with the other external actuators.

It is important to notice that our architecture accepts a wide range of data/information/knowledge sources, from sensors to persons, passing by all media sources. To illustrate this we will consider a tourist that will visit a city during some days. Our Tourism points of interest can be seen as our AmI environment. In order to decide if this tourist will start with a visit to an open air attraction or to a closed attraction the weather conditions will be important. We can obtain these weather conditions directly from sensors (temperature, humidity, etc), or by means of a friend living in this city. Notice that the granularity and quality is different according to the sources. A temperature sensor is able to inform that we have 34.6 ºC, on the other hand the tourist friend is able to say that it is too hot, but that usually it is cooler in the early morning. Another interesting aspect is that our AmI system is able to act directly on the AmI environment, or to interact with someone that will act on the environment. This means that our AmI system is able to operate like an Intelligent Agent or like a Decision Support System, or like both since some tasks can be done autonomously and other may require the intervention of human beings. The human beings, supported by the AmI system, have the responsibility to select which tasks can be done autonomously and which tasks need to pass by the human interaction. If the human beings agree with the suggestions of the decision support system in some specific tasks, there is a trend that in the future these tasks will be performed autonomously (by means of an intelligent agent or another autonomous device). The AmI system is also according to the social computing paradigm, since the interaction with the external actuators of the AmI environment or with other persons acting like external sources is allowed.

The ISyRAmI architecture considers several modules. The first is related with the acquisition of data, information and even knowledge. This data/information/knowledge deals with our AmI environment and can be acquired in different ways (from raw sensors, from the web, from experts). The second module is related with the storage, conversion and handling of the data/information/knowledge. It is understood that incorrectness, incompleteness, and uncertainty are present in the data/information/knowledge. The third module is related with the intelligent reasoning on the data/information/knowledge of our AmI environment. Here we include knowledge discovery systems, planning, multi-agent systems, simulation, optimization, etc. The last module is related with the actuation in the AmI environment, by means of automation, robots, intelligent agents and users, using decision support systems in this last case.

Figure 2 shows the four internal modules of the ISyRAmI architecture and the way in which they interact. This Figure also shows how SPARSE and the Intelligent Tutoring System (ITS) will be integrated in our AmI architecture. Both SPARSE and the ITS are spread along the last three modules of the AmI architecture. The Data/Information/Knowledge Acquisition module is able to collect all data received from the AmI Environment, the SCADA (supervisory control and data acquisition) system, and from other entities involved in the operation of the power system (CO, CTCH, CG, EDIS), presented in section 4.2. The second module, the Data/Information/Knowledge Storage, Conversion, and Handling module, includes the knowledge base of SPARSE expert system, and the knowledge bases of the tutoring tools. These knowledge bases comprise the operator's models, the knowledge about the teaching domain, and the instructional knowledge. The Intelligent Reasoning module integrates the inference engine of the expert system, the operator's modelling mechanism, and the simulation module of CoopTutor. The last module of the architecture, the Decision Support/Intelligent Actuation module, accomplishes the intelligent behaviour of the two applications integrated in the AmI environment. The SPARSE expert system act as a decision support tool, assisting control centre (CC) operators during its daily operation tasks. By the Power System grid company demand and for security reasons, the SPARSE's recommendations about restoration procedures are not executed autonomously. Therefore SPARSE does not interact directly with the AmI physical environment. The tutoring tools do not operate in the AmI physical environment. The last module of our AmI achitecture implements the intelligent behaviour of DiagTutor during a training session about incident diagnosis. In the same sense CoopTutor also does not operate directly in the physical environment. However, this training tool act on an AmI simulated environment reflecting the behaviour of the simulated actions performed by the tutor and by other simulated entities involved in the restoration procedures. These simulated entities correspond to the simulation of real entities such as CO, CTCH, CG, and EDIS. Due to the presence of an ITS in our AmI architecture, this must consider two types of AmI environments: a physical and a simulated environment.

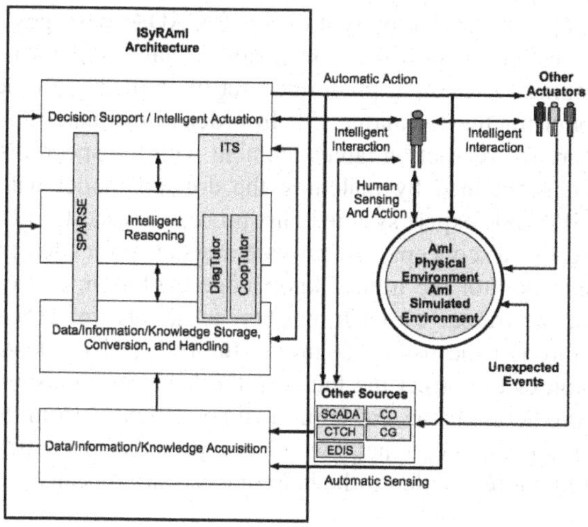

Fig. 2 ISyRAmI architecture

4 Control Centre Training Tools

SPARSE is a project initially created for helping Power Systems CC operators in the diagnosis of incidents and power restoration [4]. This system is a good example of context awareness since it is aware of the on-going situation, acting in different ways according the normal or critical state of the power system. Another project closely related with SPARSE involved the development of a training tool based in an Intelligent Tutoring approach. This training tool helps in training of new operators in diagnosis and restoration procedures of the Power system.

Presently, our aim is to integrate these two systems in a power system control room using an Ambient Intelligent framework based on the architecture presented in section 3. In this section we will present in brief the ITS included in the AmI architecture. This ITS includes two training tools: one devoted to the training of fault diagnosis skills and another dedicated to the training of power system restoration techniques.

4.1 DiagTutor

During the analysis of lists of alarm messages, CC operators must have in mind the group of messages that describes each type of fault. The same group of messages can show up in the reports of different types of faults. So CC operators have to analyze the arrival of additional information, whose presence or absence determines the final diagnosis. Operators have to deal with uncertain, incomplete and inconsistent information, due to data loss or errors occurred in the data gathering system. The interaction between the trainee and the tutor is performed through prediction tables where the operator selects a set of premises and the corresponding conclusion. The premises represent events (SCADA messages), temporal constraints between events or previous conclusions. DiagTutor does not require the operator's reasoning to follow a predefined set of steps, as in other implementations of the model tracing technique [5]. In order to evaluate this reasoning, the tutor will compare the prediction tables' content with the specific situation model [6]. This model is obtained by matching the domain model with the inference undertaken by SPARSE expert system. This process is used to: identify the errors revealing operator's misconceptions; provide assistance on each problem solving action, if needed; monitor the trainee knowledge evolution; and provide learning opportunities for the trainee to reach mastery. In the area of ITS's this goals has been achieved through the use of cognitive tutors [7]. The identified errors are used as opportunities to correct the faults in the operator's reasoning. The operator's entries in prediction tables cause immediate responses from the tutor. In case of error the operator can ask for help, which is supplied as hints. The first hints are generic, becoming more detailed if the help requests are repeated.

4.2 CoopTutor

The management of a power system involves several distinct entities, responsible for different parts of the network. The power system restoration needs a close co-ordination between generation, transmission and distribution personnel and their actions should be based on a careful planning and guided by adequate strategies. In the specific case of the Portuguese transmission network, four main entities can be identified: the National Dispatch Center (DC), responsible for the energy management and for the thermal generation; the Operational Centre (CO), controlling the transmission network; the Hydroelectric Control Centers (CTCH), responsible for the remote control of hydroelectric power plants and the Distribution Dispatch (EDIS), controlling the distribution network.

The power restoration process is conducted by these entities in such a way that the parts of the grid they are responsible for will be slowly led to their normal state, by performing the actions specified in detailed operating procedures and fulfilling the requirements defined in previously established protocols. This process requires frequent negotiation between entities, agreement on common goals to be achieved, and synchronization of the separate action plans on well-defined moments. These agents can be seen as virtual entities that possess knowledge about the domain. As real operators, they have tasks assigned to them, goals to be achieved and beliefs about the network status and others agents' activity. They work asynchronously, performing their duties simultaneously and synchronizing their activities only when this need arises. In our system, the trainee can choose to play any of the available roles, namely the CO and the DC ones, leaving to the tutor the responsibility of simulating the other participants. The representation method used to model the trainee's knowledge about the domain knowledge is a variation of the Constraint-Based Modeling technique [8]. This student model representation technique is based on the assumption that diagnostic information is not extracted from the sequence of student's actions but rather from the situation, also described as problem state that the student arrived at. Hence, the student model should not represent the student's actions but the effects of these actions [9]. This tutoring tool is able to train individual operators as if they were in a team, surrounded by virtual "operators", but is also capable of dealing with the interaction between several trainees engaged in a cooperative process. It provides specialized agents to fulfill the roles of the missing operators and, at the same time, monitors the cooperative work, stepping in when a serious imbalance is detected.

5 Conclusions and Further Work

This paper proposed an architecture for Ambient Intelligent Systems. This architecture is oriented for the intelligent behaviour of AmI environments and is divided in four modules described in the paper. The paper also represents the external parts of the system, namely the other actuators on the AmI environment, the other sources of data, information, and knowledge, the automatic sensing from the AmI environment, the automatic actuation on the AmI environment, the interaction with the user, and the unexpected events that can change the AmI environment state. The architecture proposed is based on the ISyRAmI methodology and architecture. This architecture was used to integrate two control centre applications in an AmI environment. However, since these to applications correspond to work developed before the ISyRAmI architecture development, they do not follow it explicitly. It was necessary two adapt them to the proposed architecture. Further work will include the evaluation of the integration of the tutoring system in the AmI environment in order to demonstrated the merits of the proposed architecture.

Acknowledgments. The authors would like to acknowledge COMPETE Program, and the Portuguese Science and Technology Foundation for their support to GECAD unit, and CITOPSY Project (reference PTDC/EEA-EEL/099575/2008).

References

1. Ramos, C.: An Architecture for Ambient Intelligent Environments. In: Corchado, J.M., Tapia, D.I., Bravo, J. (eds.) 3rd Symposium of Ubiquitous Computing and Ambient Intelligence, vol. 51, pp. 30–38. Springer, Heidelberg (2009)

2. Ramos, C.: Ambient Intelligence Environments, in Encyclopedia of Artificial Intelligence. In: Rabñal, J.R., Dorado, J., Sierra, A. (eds.) Information Science Reference (2008), ISBN 978-1-59904-849-9
3. Ramos, C., Augusto, J.C., Shapiro, D.: Ambient Intelligence: the next step for Artificial Intelligence. IEEE Intelligent Systems -Special Issue on Ambient Intelligence, Guest Editors' Introduction 23(2), 15–18 (2008)
4. Vale, Z., Machado Moura, A., Fernandes, M.F., Marques, A., Rosado, C., Ramos, C.: SPARSE: An Intelligent Alarm Processor and Operator Assistant for Portuguese Substations Control Centers. IEEE Expert - Intelligent Systems and Their Application 12(3), 86–93 (1997)
5. Anderson, J., Corbett, A., Koedinger, K., Pelletier, R.: Cognitive Tutors: Lessons Learned. The Journal of the Learning Sciences 4(2), 167–207 (1995)
6. Faria, L., Silva, A., Vale, Z., Marques, A.: Training Control Centers' Operators in Incident Diagnosis and Power Restoration using Intelligent Tutoring Systems. IEEE Transactions on Learning Technologies 2(2) (April-June 2009)
7. Aleven, V., Koedinger, K.R.: An effective meta-cognitive strategy: learning by doing and explaining with a computer-based Cognitive Tutor. Cognitive Science 26(2), 147–179 (2002)
8. Ohlsson, S.: Constraint-Based Student Modeling. In: Greer, McCalla (eds.) Student Modeling: the Key to Individualized Knowledge-based Instruction, pp. 167–189. Springer, Heidelberg (1993)
9. Silva, A., Vale, Z., Ramos, C.: Cooperative Training of Power Systems Restoration Techniques. In: 13rd International Conference on Intelligent Systems Applications to Power Systems, Washington, pp. 36–42 (November 2005)

Ambient Intelligence Based Architecture for Immersive Social Entertainment

Mª Amparo Navarro, Ana Belén Sánchez, Carlos Fernández,
and Mª Teresa Meneu

Abstract. The new entertainment ICT are networked but not fully integrated into the environment. Project Enjoy.IT! wants to design, develop and validate an entertainment platform with advanced contents which will set up a practical realization of the new products and services from the Future Internet. On the creation of this entertainment platform we pretend to promote the use of ICT by users which are not familiarized with this kind of technologies, such as children. Project Enjoy.IT! wants to integrate the physical world as an extension of the virtual world and vice versa. Thus, the project is going to create an AmI system which will be able to act depending on the user's necessities, and children could be immersed in a world that will interact to them without being aware of what technology are they using. The platform will be based on Services Choreography which will be a modular and scalable and it will allow an easy and simple integration of the necessary elements to give support to interactive entertainment activities.

1 Introduction

Current society is increasingly characterized by the whole presence of Information and Communications Technologies (ICT) in all fields and daily-life applications. ICT focused on entertainment for the youth are totally networked: online videogames and entertainment connecting players together have online communities associated. The new entertainment ICT are networked but not fully integrated into the environment.

The current provision of the sociocultural sphere needs a qualitative transformation that allows the full integration of ICT into the environment of the person, in an internal (psychological) and external (social) level. The point is that the different technological tools and devices can be connected and share information to each other about the individual's conditions, capabilities, emotions, reasoning, needs, choices, etc. The entertainment ICT have to be totally and properly integrated into the environment within the person that is paying attention on it. The point is that a person can freely interact without being aware of the presence of

Mª Amparo Navarro · Ana Belén Sánchez · Carlos Fernández · Mª Teresa Meneu
ITACA - Health and Wellbeing Technologies, Universidad Politécnica de Valencia
e-mail: {manasal,absanch,carfell,tmeneu}@itaca.upv.es

P. Novais et al. (Eds.): Ambient Intelligence - Software and Applications, AISC 92, pp. 227–231.
springerlink.com

ICT, and this person could develop in the technology world easily, as if these new implementations were part of the environment, as if they were there as well as the rest of the objects that compose the social reality. This is a complete adaptation of individual to ICT and ICT to the individual, a continuous feedback.

In this context Project Enjoy.IT! [1] is born, whose main objective is to design, develop and validate an entertainment platform with advanced contents which will set up a practical realization of the new products and services from the Future Internet. By creating this platform we pretend to promote the use of ICT by users which are not familiarized with this kind of technologies, such as children.

2 Materials and Methods

Performing a global vision of the video games existing nowadays, it is worth focusing on serious games. A serious game is a game designed for a different purpose other than pure entertainment. Serious games are objects and/or learning tools which have pedagogical and educational objectives that enable players to obtain a set of predominantly practical knowledge and skills. This statement is supported by the intrinsic educational characteristics of the game because it is a motivating factor for students, it gets meaningful and leisure learning, it promotes teamwork and it has great flexibility in its use [4, 5].

For this reason, what we aim to achieve with Project Enjoy.IT! video gaming platform is that children learn while they are playing, at the same that they are developing in both the psychomotor activity and the psychosocial levels, being simultaneously aware of the new Knowledge Society and the Internet of Things [2, 7].

The interaction form of these games should be different from the usual: we want children to perform a series of activities linked by a common history, in which they will have to move and interact with the environment, either through sensors responding to his actions, or using the new available game controllers that make users stand up from the chair and use their body to move the remote control and the character at the same time, such as Wiimote or PlayStation Move. Or even using Microsoft Kinect, where gestures and spoken commands are used to control the game.

Project Enjoy.IT! wants to integrate the physical world as an extension of the virtual world and vice versa. Thus, the system intelligence and the users' interaction can overcome the digital barriers. Project Enjoy.IT! is going to create an **AmI** [3] system which will be able to act depending on the user's necessities. But this AmI network will not only be able to realize received requests but also be aware of the user's identity and his context, customs and circumstances.

Project Enjoy.IT! wants to integrate all this technologies into the world of the socio-cultural animation, which focuses on the development of recreational activities for children of all ages. This animation currently does not include ICT for the realization of the games and that is where Project Enjoy.IT! is going to give a change with this type of combined experience.

3 Results

Project Enjoy.IT! aims to create a digital and interactive entertainment platform for children. A complete role-playing adventure of interconnected games is being built. Ten groups of twenty children (9-10 years old) from the Polytechnic University of Valencia's Summer School will participate on it. Each group will play a village from a region. Natural resources are near extinction and children have to begin a journey around an imaginary world in order to find several pieces from a magical object which will solve the problem. Using the recovered piece by each group the magical object will be assembled, always bearing in mind the collaborative spirit to promote positive values among children. Each game will be multiplayer, collaborative (among children and different groups) and interconnected to the whole adventure, using a Social Network. All the advances of each group could be consulted by every players. For this reason, a social network will be created where users will participate either individually or forming groups by their roles, with contents both generated by them and automatically by the system from the information captured by the sensors and the intelligence of the platform, supplying information about the evolution of activities and becoming another useful element for the realization of activities, strengthening interaction among users, corporative strategies and social relations.

The project will take up the challenge of supplying a modular and scalable platform that allows an easy and simple integration of the necessary elements to give support to interactive entertainment activities.

The results of the project will be the development of new tools for the creation and use of digital contents on a platform combining multiple sensor devices and wired and wireless communication technologies that allow the establishment of a link between the physical and virtual worlds. At the same time, all that will be merged with the emergent social networks and will apply pedagogical methodologies.

To achieve all the above the following architecture is proposed, which is based on a Process Choreographer (see Figure 1). Using the Process Choreographer, services and devices are able to exchange information in a distributed way. This means that choreographed processes are independent and can communicate with each other to define its workflow execution. This model makes it easy to turn the connection and disconnection of services dynamically and is capable of using different types of sensors and configurations [6].

Use of choreography for interconnecting services requires the use of a common exchange language that allows services to be understood. This is achieved with an architecture that includes a semantic layer in the choreographer to improve communication among sensors, actuators, and services in the system. The use of Ontology Services and Machine Reasoning for the description of data from sensors allows to make a more accurate interpretation of information obtained from them, and allows the system to automatically detect sensors and services available in each moment [6].

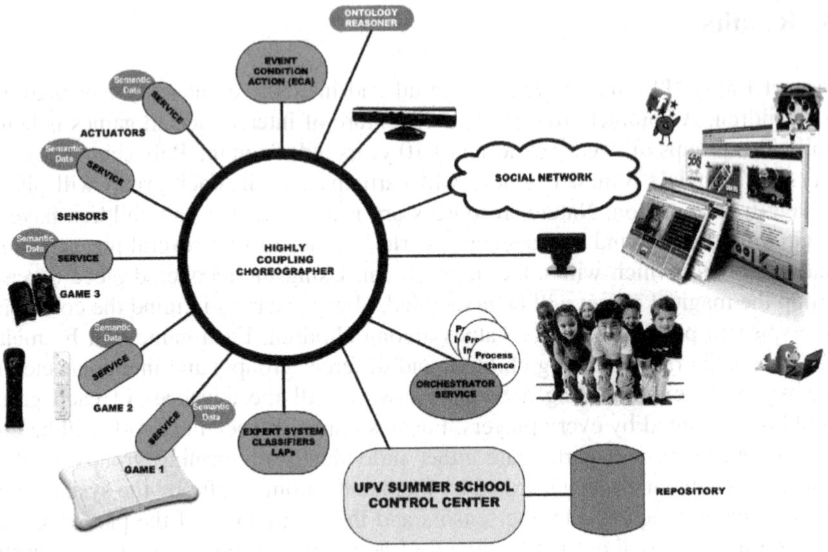

Fig. 1 Proposed platform architecture

It is also included in the architecture an Orchestrator of Services connected to the Choreographer who supports the use of Workflows to describe processes graphically; characterized by being a formal specification of the implementation of processes that can be executed dynamically using a language called Workflow Engine.

Implementation of an individualized model of the user's behaviour is a great technological complexity task. The aim is to turn that model on a Workflow for automatic processing.

It is noteworthy that it will exist a control centre where the evolution of the context could be watched in a centrally way. This will be made interviewing each part of the system. There is also a repository where all data items will be stored as well as the progress of each group.

4 Discussion and Conclusions

The technological platform proposed is projected to make technology accessible to children, to provide personal and social benefits arising from the use of ICT, and to contribute to the development of a significant set of values, beliefs, thoughts and attitudes as well as to the development of several physical and cognitive skills. To integrate ICT in the individual's environment will allow to change the contexts of interaction, to reach the inner child, defining and mediating her personal experience. The purpose is to bring the ways in which children define their own world so that ICT can be properly integrated.

The aim is to adapt personal and social experiences of children in order to provide benefits to them. It is sure that the new entertainment ICT and the platform in which we are working, besides being a primary cognitive tool in the sense that the players enhance mental skills, and allow the development of new ways of thinking, will also contribute to promote social relationships, cooperative learning, the development of capacities for creativity, communication and reasoning and other essential components of the interaction and the physical and psychosocial growth. The AmI system will immerse the users (children) in a world that will interact to them without being aware of what technology are they using.

The first step will be a pilot on July 2011, where new videogames using AmI will be tested by several groups of twenty children ten years old. The second step will be the test of the whole platform on July 2012 by two hundred children. Future researches will include different ages of children by including new games as well as the ability to set the difficulty of the games according to age.

References

1. Project Enjoy.IT, `http://www.proyectoenjoyit.es/` (accessed October 2010)
2. International Telecommunication Union. ITU Internet Reports, The Internet of Things (2005)
3. Riva, G., Vatalaro, F., Davide, F., Alcañiz, M.: Ambient Intelligence. IOS Press, Amsterdam (2005)
4. Sánchez, M.: Buenas Prácticas en la Creación de Serious Games (Objetos de Aprendizaje Reutilizables). In: Post-Proceedings del IV Simposio Pluridisciplinar sobre Diseño, Evaluación y Desarrollo de Contenidos Educativos Reutilizables, SPDECE 2007, Bilbao, Spain, September 19-21 (2008)
5. Ritterfeld, U., Cody, M.J., Vorderer, P.: Serious Games. Mechanisms and effects. Routledge, New York (2009)
6. Fernández, C., Lázaro, J.P., Benedí, J.M.: Workflow Mining Application to Ambient Intelligence Behavior Modeling. In: Stephanidis, C. (ed.) UAHCI 2009. LNCS, vol. 5615, pp. 160–167. Springer, Heidelberg (2009)
7. Sundmaeker, H., Guillemin, P., Friess, P., Woelfflé, S.: Vision and challenges for realising the Internet of Things. In: Cluster of European Research Projects on the Internet of Things (2010)
8. Fernández, C., Mocholí, J.B., Moyano, A., Meneu, T.: Semantic Process Choreography for Distributed Sensor Management. In: International Workshop on Semantic Sensor Web - SSW 2010 (2010)

Grouping Behaviour in AmI-Enabled Crowd Evacuation

Alexei Sharpanskykh and Kashif Zia

Abstract. Grouping behaviour occurs often in crowd evacuation. On the one hand, groups are needed for efficient evacuation. On the other hand, large uncontrolled groups (herds) may cause clogging and increase panic. The mechanisms of emergence of leaders and groups in complex socio-technical systems with intelligent technical components are not well understood. This paper presents the first attempt to unveil the role of AmI technology in formation of spontaneous groups in crowd evacuation. To this end several hypotheses were formulated, which were tested by simulation experiments based on a cognitive agent model. The checking of the hypotheses was done in the context of a train station evacuation scenario. The general outcome is that in a system with scarce and uncertain information, AmI technology can be used to stimulate emergence of leaders and groups to increase the efficiency of evacuation. Furthermore, a large penetration rate of ambient devices may be unnecessary and even not appropriate for fluent evacuation.

1 Introduction

In the literature [1, 9, 10, 13] it is indicated that people often form spontaneous groups during evacuation. On the one hand, dynamic formation of groups is recognised as a prerequisite for efficient evacuation [1,13]. On the other hand, large uncontrolled groups, sometimes called herds, may cause clogging of paths and increase panic [1, 9]. In examples of efficient evacuation emergent leaders played a prominent role in guiding of and sustaining a steady emotional state in groups [1]. In social psychology several sources of power of informal (or emergent) leaders are recognised, among which knowledge and physical traits are the most essential ones [2]. AmI technology can be used to discover and propagate knowledge in socio-technical systems. AmI-equipped actors may obtain scarce, and thus valuable, knowledge about the environment serving as one of the important sources of emergent leadership. Although much research has been done on emergent

Alexei Sharpanskykh
VU University Amsterdam, De Boelelaan 1081a, Amsterdam, The Netherlands

Kashif Zia
Institute for Pervasive Computing, Johannes Kepler University, Linz, Austria

P. Novais et al. (Eds.): Ambient Intelligence - Software and Applications, AISC 92, pp. 233–240.
springerlink.com

leadership in social systems [2], the mechanisms of emergence of leaders and groups in complex socio-technical systems with intelligent technical components are less clear. This paper presents the first attempt to unveil the role of AmI-technology in formation of spontaneous groups in crowd evacuation. The specific research question addressed in this paper: *how does AmI technology influence the emergence of groups and leaders in socio-technical systems with scarce and uncertain information?* For this question several hypotheses were formulated. **Hypothesis 1:** *AmI-equipped humans, who obtain up-to-date information about the environment, are recognised as leaders in emergent groups in organisations with scarce and uncertain information.* Previous studies showed that in general humans have a loyal attitude to information provided by (intelligent) technology. The validation of this statement for crowd evacuation is necessary, but also problematic, as such experiments cannot be organised easily. To address this issue, we examined three conditions: (a) humans have high initial trust to technology and distrust it slowly after negative experiences; (b) humans trust technology in the same manner as to human strangers; (c) humans have high initial trust to technology (initial bias), but distrust it rapidly after negative experiences. Note that trust to technology is dynamic and depends on the human's experiences with technology. In relation to these conditions the following hypotheses are formulated: **Hypothesis 2:** *More grouping behaviour is observed under (a) and (c) conditions than under (b) condition.* **Hypothesis 3:** *Humans under (c) change groups more frequently than under (a).* To quantify these hypotheses the measures *following index, change index* and *group size* are introduced in section 4. One more hypothesis to be tested in the paper is related to *the large group effect* known for social emergency systems [1]: **Hypothesis 4:** *Evacuation with larger groups proceeds more slowly (less efficiently) than with smaller groups.* The hypotheses were tested in the frames of an emergency case study introduced in Section 2. For formal verification of the hypotheses a cognitive agent-based model was developed, in which humans and AmI devices were represented by agents. This model is based on a number of theories from Neuropsychology, Social Science and Psychology, many of which were empirically validated. The model is described in Section 3.The verification results for the hypotheses are presented in Section 4. Section 5 concludes the paper.

2 Case Study

Since it is nearly impossible to perform an evacuation trial to validate an emergency egress strategy at a mass place, an agent-based social simulation approach was taken instead. In this simulation study we focussed on evacuation of a train station. To ensure that the simulation setting is a true representative of reality, we incorporated real CAD design of an existing Austrian main railway station to generate the space along with observed population statistics. The station in the simulation model had 3 exits with different flow capacities. The station was populated randomly with 500 agents representing humans, from which 10 agents were equipped with AmI technology. The AmI technology in focus is the LifeBelt, specially designed for emergency situations [4]. Particularly, it exploits the unused information transmission capability of sense of touch instead of usual visual and auditory perception to deliver the message. In this way, it does not deviate and frustrate the human already overwhelmed with the visual and auditory perceptual overload. The LifeBelt system exploits the position and variation of vibro-tactile stimuli to indicate both orientation (intended direction) as well as urgency (intended speed) through tactor-elements embedded into a hip worn belt. Eight vibrator elements are lined up in the fabric of a hip belt, and connected to the belt controller. The controller activates the vibrator switches according to commands received wirelessly from a global control unit, in this case a global 'evacuation control unit'. We assume that the global control unit provides reliable, up-to-date information to all LifeBelts without any noise. Only few individuals are considered to have worn the LifeBelt who would act as leaders during evacuation. The recommendation for an exit is generated by evacuation control unit under the influence of exit area dynamics (e.g. flow, density), which are assumed to be measured by a technology mounted on the exit. The exit choice would then be communicated to all the LifeBelts. Each LifeBelt has a location map used to transform the coordinates of an exit to the desired orientation to move. Thus, AmI equipped agents have direct access to information essential for successful evacuation, which they could propagate further by interaction with other agents. Agents interact with each other *non-verbally* by spreading emotions and intentions to choose particular exits, and *verbally* by communicating information about the states of the exits. Since the agents with LifeBelts possess information about the exits not available to the agents without LifeBelts, the AmI-equipped agents hold one of the most important sources of power identified in social studies on emergent leadership [2]. However, the agents without devices are still free to decide whether to follow AmI-equipped agents or to rely on their own beliefs and exit choices. It is important to stress that the grouping effect is not encoded in our model explicitly, but emerges as a result of complex decision making by agents. The purpose of this study is to determine to which extent agents without AmI devices would follow AmI-equipped agents in the process of evacuation.

3 Cognitive Agent Model

To model decision making of agents representing humans a general affective decision making model from [11] was instantiated for the case study. The model is formalised using a temporal state transition system format [12]. Depending on a situational context an agent determines a set of applicable options to satisfy its goal. In the case study the goal of each agent is to get outside of the building in the fast possible way. This is achieved by an agent by moving towards the exit that provides for fastest evacuation as it perceived by the agent. Evacuation options are represented internally in agents by one-step simulated behavioural chains, based on the neurological theory by Hesslow [5] (see Fig.1). In Fig. 1 the burning station situation elicits activation of the state srs(evacuation_required) in the agent's sensory cortex that leads to preparation for action preparation_for(move_to(E)). Here E is one of the exits of the station. Note that if more than one exit is known to the agent, then in each option representation the preparation state corresponding to the option's exit is generated. Then, associations are used such that preparation_for(move_to(E)) will generate srs(is_at(E)), which is the most connected sensory consequence of the action move_to(E).

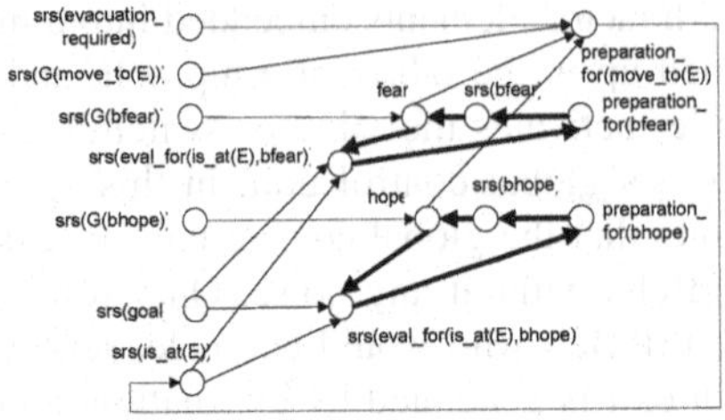

Fig. 1 The emotional decision making model for the option to move to exit E

The strength of the link between a preparation for an action and a sensory representation of the effect of the action (see Fig.1) is used to represent the confidence value of the agent's belief that the action leads to the effect. The simulated sensory states elicit emotions, which guide agent behaviour either by reinforcing or punishing simulated actions. By evaluating sensory consequences of actions in simulated behavioural chains using cognitive structures from the OCC model [8], different types of emotions can be distinguished. In the example two types of emotions - fear and hope – are distinguished, which are often considered in the emergency domain. According to [8], the intensity of fear induced by an event depends on the degree to which the event is undesirable and on the likelihood of the event. The intensity of hope induced by an event depends on the degree to which the event is desirable and on the likelihood of the event. Thus, both emotions are generated based on the evaluation of a distance between the effect states for the action from an option and the agent's goal state. In particular, the evaluation function for hope in the evacuation scenario is specified as *eval(g, is_at(E))* = ω, where ω is the confidence value for the belief about the accessibility of exit E, which is an aggregate of the agent's estimation of the distance to the exit and the degree of clogging of the exit. Although it is assumed that the distances to the exits are known to the agents, the information about the degree of clogging of the exits is known only to AmI-equipped agents. Emotions emerge and develop in dynamics of reciprocal relations between cognitive and body states of a human [3]. These relations, omitted in the OCC model, are modelled from a neurological perspective using Damasio's principles of 'as-if body' loops and somatic marking [3]. The as-if body loops for hope and fear emotions are depicted in Fig. 1 by thick solid arrows. The following rules describe the evolution of the emotional states:

srs(eval_for(is_at(E),bhope),V2) & srs(G(bhope),V1)
$\twoheadrightarrow$ hope(o, (β_h- β_h *(1-V1)*(1-a1) + (1-β_h)*V1*a2)/(1- β_h*(1-V1)*a1 - (1-β_h)*V1*a1)),
where a1 = β_h - 2*β_h*V2 + V2, a2 = β_h – β_h*(1-V2)
srs(eval_for(is_at(E),bfear),V2) & srs(G(bfear),V1)
$\twoheadrightarrow$ fear(o, (β_f - β_f *(1-V1)*(1-a3) + (1-β_f)*V1*a4)/ (1- β_f*(1-V1)*a3 - (1-β_f)*V1*a3)),
where a3 = β_f*V2+1-V2-β_f*(1-V2), a4 = β_f – β_f*V2

here β_h is the degree of extraversion (i.e., tendency to experience positive emotions) of the agent; β_f is the degree of neuroticism (i.e., a tendency to experience negative emotions) of the agent; G(bhope) is the aggregated preparation to the emotional response (body state) of the agent's social neighbourhood. The social influence on the individual decision making is modelled based on *the mirroring function* [6] of preparation neurons in humans. Such neurons, in the context of the neural circuits in which they are embedded, show both a function to prepare for certain actions or bodily changes and a function to mirror similar states of other persons. This mirroring function in social decision making is realised in two forms: (1) by *mirroring of emotions*, which indicates how emotional responses in different agents about a decision option mutually affect each other, and (2) by *mirroring of intentions* or *action preparations* of individuals for a decision option. Furthermore, the social influence includes spread of beliefs of agents supporting or prohibiting options (e.g., the belief about the accessibility of an exit).

The mirroring is realised through information and emotion contagion processes. The contagion strength of the interaction from agent B to agent A is defined as follows: $\gamma_{BA}=\varepsilon_B \cdot \text{trust}(A, B) \cdot \delta_A$, here ε_B is the personal characteristic expressiveness of the sender (agent B), δ_A is the personal characteristic openness of the receiver (agent A). *Trust* is an attitude of an agent towards an information source that determines the extent to which information received by the agent from the source influences agent's belief(s). The trust to a source builds up over time based on the agent's experience with the source. In particular, when the agent has a positive (negative) experience with the source, the agent's trust to the source increases (decreases). Currently experiences are restricted to information experiences only. An information experience with a source is evaluated by comparing the information provided by the source with the agent's beliefs about the content of the information provided. The experience is evaluated as positive (negative), when the information provided by the source is confirmed by (disagree with) the agent's beliefs. The following property describes the update of trust of agent A to agent B based on information communicated by B to A about the degree of contagion around exit e:

trust(A_i, A_j, V1) & communicated_from_to(A_j, A_i, congestion(e, V2)) & belief(A_i, congestion(e, V3)) $\rightarrow$ trust(i, j, V1+ γ_{tr}*(V3/(1 + e^{α}) – V1)),
here α=-ω1*(1-|V2-V3|) + 4.

According to the Somatic Marker Hypothesis [3], each represented decision option induces (via an emotional response) a feeling which is used to mark the option. For example, a strongly positive somatic marker linked to a particular option occurs as a strongly positive feeling for that option. To realise the somatic marker hypothesis in behavioural chains, emotional influences on the preparation state for an action are defined as shown in Fig. 1. Through these connections emotions influence the agent's readiness to choose the option.

4 Experiments and Results

The model was implemented in the Netlogo simulation tool [14] by cellular automata. In this tool the environment is represented by a set of connected cells, where moveable agents (turtles) reside. Cells can be walkable (open space and exits) and not-walkable (concrete, partitions, walls). Each cell of the environment is accessible from all the exits. Based on the conditions (a)-(c) identified in the introduction, three simulation settings S1-S3 were determined (see Table 1, the upper part). To test the hypothesis 4, setting S4 was identified, in which AmI-equipped agents were able to propagate information in a large range. Since the model contains stochastic elements, 10 trials were performed for each simulation setting with 500 agents. To evaluate the hypotheses three evaluation metrics were introduced: *following index (fi)*, which reflects the degree of following of AmI-equipped agents by other agents, *change index (ci)*, reflecting the frequency of group change by agents, and *group size (gs)*. As shown below, the metrics are defined per an AmI-enabled agent L (i.e, fi_L, ci_L, gs_L) and by taking the mean over all AmI-equipped agents (i.e., fi, ci, gs). The following index is defined as follows:

$fi_L = 1/|N| \cdot \sum_{A \in N} |F_{A,L}|/(t\_last - t\_first_A)$, $\quad fi = \sum_{i \in LEAD} fi_i/|LEAD|$,
where t_first_A is such that
$\exists$o1:INFO at(communicated_from_to(L, A, inform, o1), t_first_A) & $\forall$t:TIME, o:INFO t < t_first_A & $\neg$at(communicated_from_to(L, A, inform, o), t);
$N = \{a \mid t\_first_A$ is defined$\}$; $F_{A,L} = \{ t \mid t \geq t\_first_A$ & $\exists$d1,d2: DECISION at(has_preference_for(A, d1), t) & at(has_preference_for(L, d2), t) & d1=d2 & at(distance_between(A, L) < dist_threshold, t) },
t_last is the time point when L is evacuated, LEAD is the set of all technology-equipped agents, |LEAD|=10 in all experiments.
The change index is defined as follows:
$ci_L = 1/|N| \sum_{A \in N} |S_{A,L}|$, $\quad ci = \sum_{i \in LEAD} ci_i/|LEAD|$,
where $S_{A,L} = \{ t \mid (t \in F_{A,L}$ & $(t+1) \notin F_{A,L})$ OR $((t+1) \in F_{A,L}$ & $t \notin F_{A,L}) \}$.
The group size is defined as follows:
$gs_L = \sum_{t=1..t\_last} FT_{L,t}/t\_last$, $\quad gs = \sum_{i \in LEAD} gs_i/|LEAD|$,
where $FT_{L,t} = \{ ag \mid t \geq t\_first_{ag}$ & $\exists$d1,d2: DECISION at(has_preference_for(ag, d1), t) & at(has_preference_for(L, d2), t) & d1=d2 & at(distance_between(A, L) < dist_threshold, t) }.

The obtained results are summarised in Table 1 (in the lower part). As one can see from the table, the emergence of groups with AmI equipped agents as guiding leaders occurs in all settings (fi > 0), thus, the hypothesis 1 is confirmed. The high standard deviation values for fi and gs in S2 indicate that in some trials persistent groups emerged, whereas in other trials almost no grouping occurred. This is in contrast to the other simulation settings, in which notable grouping behaviour emerged in every trial. The highest fi is observed in settings S1 and S4, in which the agents were biased positively towards technology. Hypothesis 2 is also confirmed, as the fi's for settings S1 and S3 are significantly higher than fi for setting S2. From the comparison of the ci's for settings S1 and S3 the confirmation of hypothesis 3 follows. For the hypothesis 4, first it can be observed in table 1 that the groups formed in S4 are in average 3 times larger than the groups formed in S1. Note that the only distinction between S1 and S4 is the interaction range (penetration rate) of the AmI-enabled agents. Thus, settings S1 and S4 are adequate for checking hypothesis 4. As can be seen from the table, the overall evacuation time for S1 is lower than for S4. Thus, hypothesis 4 is confirmed as well. Also, as can be seen from the results, AmI-enabled evacuation with relatively small groups (settings S1 and S3) proceeds the fastest.

Table 1 The parameters used in the simulation settings S1-S4 (the upper part) and the corresponding results for 10 simulation trials for each setting (lower part)

Simulation setting (Parameter)	S1	S2	S3	S4
Initial trust value to an AmI-enabled agent	0.9	0.1	0.9	0.9
Initial trust value to an agent without AmI	0.1	0.1	0.1	0.1
$\omega 1$ in the update of trust to an AmI-enabled agent	39	9	9	39
$\omega 1$ in the update of trust to an agent without AmI	9	9	9	9
Interaction range (in cells)	10	10	10	25
Evaluation metrics	**S1**	**S2**	**S3**	**S4**
Mean overall evacuation time (standard deviation)	147.7 (10.7)	174.4 (16.9)	150.1 (9.7)	170.3 (21.3)
Mean following index fi (standard deviation)	0.46 (0.09)	0.27 (0.12)	0.43 (0,07)	0.5 (0.09)
Mean change index ci (standard deviation)	0.48 (0.05)	0.25 (0.07)	0.92 (0.21)	0.17 (0.03)
Mean group size gs (standard deviation)	26 (8.7)	29 (19.8)	27 (8.3)	81 (14.2)

5 Conclusions

Dynamic formation of groups and emergence of leaders have a significant impact on the efficiency of evacuation [1]. However, governing principles behind these phenomena in socio-technical systems are not clearly understood. In this paper we made the first step towards understanding how AmI technology influences grouping behaviour in large-scale socio-technical systems. For this, four hypotheses were formulated, a cognitive agent model was developed, and agent-based social simulation tools were used to verify the hypotheses. Although the obtained results still require empirical validation, some of them correlate well with findings from Social Science [1, 13]. Furthermore, the simulation model developed relies strongly on a theoretical basis comprising theories from Social Science, Psychology and Neuropsychology, many of which were empirically validated. Previously, grouping (or herding) behaviour of humans in evacuation was modelled using diverse computational techniques [7, 9, 10]. However, this work largely ignores the (intelligent) technological component. Also, the human behaviour is modelled in a very simplistic way, often using classical contagion models or lattice gas principles. In the future, in collaboration with social psychologists more realistic mechanisms of emergent leadership (e.g., physiological and behavioural cues) and group formation will be integrated in the existing model.

References

1. Barton, A.H.: Communities in disaster: A sociological analysis of collective stress situations. Doubleday, Garden City (1969)
2. Clegg, S.R.: Frameworks of Power. Sage, Thousand Oaks (1989)

3. Damasio, A.: The Feeling of What Happens. Body and Emotion in the Making of Consciousness. Harcourt Brace, New York (1999)
4. Ferscha, A., Zia, K.: LifeBelt: Crowd Evacuation Based on Vibro-Tactile Guidance. IEEE Pervasive Computing 9(4), 33–42 (2010)
5. Hesslow, G.: Conscious thought as simulation of behaviour and perception. Trends in Cog. Sci. 6, 242–247 (2002)
6. Iacoboni, M.: Mirroring People: the New Science of How We Connect with Others. Farrar, Straus & Giroux (2008)
7. Isobe, M., Helbing, D., Nagatani, T.: Many Particle Simulation of the Evacuation Process from a Room Without Visibility (2003), `http://arXiv.org/abs/condmat/0306136`
8. Ortony, A., Clore, G.L., Collins, A.: The Cognitive Structure of Emotions. Cambridge University Press, Cambridge (1988)
9. Saloma, C., Perez, G.J.: Herding in real escape panic. In: Proceedings of the 3rd International Conference on Pedestrian and Evacuation Dynamics. Springer, Heidelberg (2006)
10. Sharma, S.: Avatarsim: A multi-agent system for emergency evacuation simulation. Journal of Computational Methods in Science and Engineering 9(1,2), 13–22 (2009)
11. Sharpanskykh, A., Treur, J.: Adaptive Modelling of Social Decision Making by Affective Agents Integrating Simulated Behaviour and Perception Chains. In: Pan, J.-S., Chen, S.-M., Nguyen, N.T. (eds.) ICCCI 2010. LNCS, vol. 6421, pp. 284–295. Springer, Heidelberg (2010)
12. Sharpanskykh, A., Treur, J.: A Temporal Trace Language for Formal Modelling and Analysis of Agent Systems. In: Specification and Verification of Multi-Agent Systems (book chapter), Springer, Heidelberg (2010)
13. Svenson, O., Maule, A.J. (eds.): Time pressure and stress in human judgment and decision-making. Plenum, New York (1993)
14. Web reference, `http://ccl.northwestern.edu/netlogo` (last accessed on November 2010)

A Mobile System to Visualize Patterns of Everyday Life

Nuno Correia, Armanda Rodrigues, Tiago Amorim, Jared Hawkey, and Sofia Oliveira

Abstract. The paper describes an ambient intelligence proposal, an application which runs on mobile phones, presenting individual users with an overview of their time usage patterns. It tracks users' movements in space using GPS technology and displays this information, aiming to highlight changes to their normal routine. The output, a data visualization interface, displayed on the phone is designed to produce a highly personal topology of time usage. Rooted in a deep interest in unveiling the hidden patterns of everyday life, this work aims to research Ubicomp in a personal context. By placing the software literally in the user's pocket (mobile phone), it is implied the best way for one to reflect upon the issue is through experience.

1 Introduction

Most research to date, especially in mobile and ubiquitous computing, has focused on enabling the seamless use of computers, tending to assume a passive role for the user. Rooted in a deep interest in unveiling the hidden patterns of everyday life, the work presented in this paper aims to research Ubicomp in a personal context. Our goal is to build a piece of software, which will run on users' mobile phones, tracking their movements in space, using GPS technology, and display meaningful visualizations of their data. The output, a data visualization interface, displayed on the phone is designed to produce a highly personal topology of time usage [2]. Important research questions addressed by the project and considered in the application described in this paper include: (1) How can a data visualization interface deploying an artistic approach, have a contribution to make in understanding the role of experience in relation to the real? (2) How can data

Nuno Correia · Armanda Rodrigues · Tiago Amorim
CITI and DI/FCT/UNL, Quinta da Torre, 2829 -516, Caparica, Portugal
e-mail: {nmc,arodrigues}@di.fct.unl.pt

Jared Hawkey · Sofia Oliveira
CADA, Ed. Interpress, Rua Luz Soriano, 67, 3°, sala 43, 1200 – 246, Lisbon, Portugal
e-mail: jaredhawkey@gmail.com, sofiaoliveira@cada1.net

P. Novais et al. (Eds.): Ambient Intelligence - Software and Applications, AISC 92, pp. 241–245.
springerlink.com

visualization techniques be improved to create readable and playful images, which also require interpretation, while allowing for user input and stimulating reflection?

(3) How will it work locally on the mobile phone? Our goal is to explore the phone itself as a worthwhile object of investigation: to exploit the computing and potential of a device held by 4,6 billion people.

Though interested in patterns of everyday life, this project does not aim to generate space-behavior profiles of mobile users. The focus is on the individual, and patterns are seen as something which might be interesting to reflect upon. The potentialities and limitations of data capture and mining will be tested in real-life situations with real users, thus aiming to research novel means for improved sensitivity to personal idiosyncrasies and individual needs.

A relevant approach to this project is Dourish's critique of Ubicomp's cognitive, task-driven approach to HCI (Human Computer Interaction) [1]. The need to represent and study dynamic spatial data is most commonly instantiated by the need for sophisticated representations of spatio-temporal data (e.g, the Space-Time Cube, which has been extended in recent work [4]). This happened in different disciplines, even to highlight a non-spatial dimension, like in Ringmaps [5].

The existence of technology to collect and analyze massive amounts of data, produced, on a daily basis, by the activities of people and organizations, led to the realization that it is possible to identify patterns in what we do physically and electronically. The map metaphor is being extended to non-spatial contexts enabling the creation of geographic visualizations that are actually more than that (e.g., the Morphing City and Empires Decline[1]). Here the main challenge is how to create such visualizations on a mobile device. With small screens, the restrictions of the viewing window must be addressed by providing context to the user [3].

2 Visualization Proposals

The visualizations were first tested using a desktop application and combine the captured variables – location, time, duration, and frequency – in a single comprehensive interface as described below:

(1) Physical coordinates: in this visualization, each location is positioned according to a physical geographic coordinate. A southern location, is seen in the lower part of the visualization window, while a northern one in the upper part. Each circle represents a place where one spends time; a circle's diameter relates to the total time spent (the longer, the larger the circle). A circle's color intensity relates to frequency, that is, places visited more often appear darker, those less often, lighter, - accordingly events outside one's pattern appear brighter;

(2) Concentric locations: this visualization compares the frequency (number of visits to a location/time unit) associated with each location, Circles' size and

---

1 http://mondeguinho.com/master/category/information-visualization, accessed Nov 15 2010.

color have the same meaning as described above, but here the circles are ordered in a concentric fashion according to their relative size (based on duration) with smaller ones drawn over larger ones;

(3) Distances based on frequency: this visualization shows those locations which are part of one's routine and those that lie outside such a routine. Frequency is mapped along a hidden diagonal axis, bottom-left corner to top-right corner. Places visited less frequently appear to the lower left and those more often visited towards the upper right. The respective distances between visited locations, in the visualization, are also based on frequency;

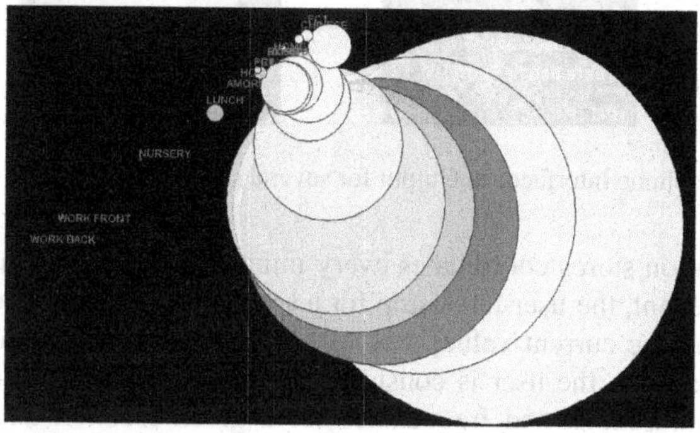

Fig. 1 Visualization with distances based on frequency

The chosen color palette - hue and tone - was designed to highlight unusual events: places more frequently visited are darker, and closer the background color, whereas those visited less often, appear relatively lighter and brighter.

3 Mobile Application

The mobile application described above runs on a mobile phone, captures GPS logs, and displays visualizations based on these logs. It was developed using J2ME (Java 2 Micro Edition) and tested with a combination of the simulator (Java Wireless Toolkit for CDLC) and Nokia N95 and N97 phones. Work has started on a new Android version.

The current mobile version includes the first, Physical Coordinates, visualization described above. Users may apply the visualization to the day they're in or a set of days, by introducing a start and finish date. Although, the other visualization proposals will be explored in the next version, an additional visualization model based on the frequency and applicable to the current day was also used at this stage. In this mobile version, it was felt, since the user carries the device they would expect to see their time usage on the interface in real time. Thus the day the user is actually in has more visual weight and only those locations visited on the day one is in are visible, positioned according to their physical coordinates. Frequency shown with color relates to the user's past movements (movements made during the previous days, depending on how it was configured by the user, -- otherwise with a default value of 3). Thus the user may compare their present movements, on the day, with their previous routines, on days before.

Fig. 2 a) Mobile phone interface; b) Output for several days using SVG

The application stores coordinates every minute but, for a given location to be considered relevant, the user must stop for a set amount of time. Based on empirical observation, the current value, was set at 3 minutes. Should the stop time be below this threshold, the user is considered to be in motion. The application accumulates the durations and frequencies for a given location and produces the outputs described above. Alongside the coordinate files, the application also generates two additional file types. First, SVG (Scalable Vector Graphics) is used to generate resolution independent images, capable of being printed or viewed on screens larger than those of the average mobile devices. Second, KML (Keyhole Markup Language) versions of the collected data, an XML format used to export to Google Earth. Presently, the KML files are mainly used as a debugging tool, to assess the validity of the location capture mechanisms but this could also be used as an alternative visualization mechanism, using platforms such as Google Earth.

4 Evaluation

The mobile application is being evaluated in several different ways. The members of the project team use the application in their daily routines and help to debug the current version. Alongside this ongoing process, several users have tried the application's more specific features, namely the visualizations. Tests were made by ten users using Nokia N79 phones, with the request to use the application during four days, after which they were asked to generate SVG and KML files to assess the validity of the results. Before using the application, users were shown a visualization and asked questions about its meaning. While all users responded correctly to the questions, and feedback was generally positive some felt the color scale could be improved, suggesting the user would be benefited by being allowed to define and associate colors for the places they visited more or less frequently.

Having used the application, users were given another questionnaire to evaluate its ease of use and whether the final results corresponded their actual routines. In general, users found it easy to understand the visualization for the day they were in, but were less positive about visualizations for more than one day. To check if the geographic coordinates corresponded to what users' regarded as meaningful, KML files were generated and the results scrutinized by the testers, using Google Earth. When asked if the locations seen in Google Earth corresponded to actual places in their lives, 56% of the testers answered 5, 33% 4 and 11% 3, on a 5-point Likert scale. Users also suggested locations should be aggregated so that a given meaningful location, rather than being represented by several coordinates, would be more clearly identifiable as one place.

5 Conclusions and Future Work

Included within the Time Machine project, the application described presents a first attempt to visualize individual time-usage patterns while highlighting changes in established routines, which also affords the user a map of how they spend time. Defined in an artistic context, it is an ongoing ambient intelligence proposal, which will be further explored and fine-tuned in future versions of the software. As the user tests showed, data filtering is required because many different coordinates may be captured for a single meaningful place. To address this problem, locations are being clustered, using distance and time spent as variables. Other types of processing, currently being employed include the clustering and classification of days in order to find routines and respective deviations. In brief, future versions of this application, will aim to examine how individual time-usage patterns and our perceptions of them might be transformed by a novel approach to data visualization. Specifically, to explore the multitude of ways through which an individual's use and sense of time might be read, interpreted, mapped and understood using technology.

Acknowledgments. We thank DGArtes (The Directorate-General for the Arts of the Portuguese Ministry of Culture) and FCT/MCTES (Fundação para a Ciência e a Tecnologia), Portugal, for funding this research as part of the Time Machine project (PTDC/EAT-AVP/105384/2008).

References

1. Dourish, P.: What We Talk About When We Talk About Context. Personal and Ubiquitous Computing 8(1), 19–30 (2004)
2. Kraak, M.: Visualization of the Time Dimension. Time in GIS: Issues in Spatiotemporal Modelling 47, 27–35 (2000)
3. Reichenbacher, T.: Adaptive Concepts for a Mobile Cartography. Journal of Geographical Sciences 11, 43–53 (2001)
4. Vrotsou, K., Forsell, C., Cooper, M.: 2D and 3D Representations for Feature Recognition in Time Geographical Diary Data. Information Visualization, Advance Online Publication (2009), doi:10.1057/ivs.2009.30
5. Zhao, J., Forer, P., Harvey, A.: Activities, Ringmaps and Geo-visualization of Large Human Movement Fields. Information Visualization 7, 198–209 (2008)

Human Behaviour Modelling Approach for Intention Recognition in Ambient Assisted Living

Kristina Yordanova

Abstract. Human behaviour models play important role in ambient assisted living in cases where the user's long term intention is inferred from her current actions. However, at present there exist plenty of models that deal with activities form different domains, but no efforts are made to derive a uniform human behaviour modelling (HBM) approach that is suitable for various activity domains. Here we describe the first steps in the definition of such generic modelling approach, which consist of identifying a set of uniform HBM requirements and deriving a set of atomic action constants. We believe that a uniform approach can increase the model functionality, and thus lead to more effective and accurate intention recognition.

Keywords: human behaviour modelling, atomic actions, intention recognition.

1 Introduction

The understanding of human behaviour plays an important role in systems dealing with ambient assisted living where intelligent appliances strive to infer human actions and intentions in order to offer better interaction and assistance.

At present, there are plenty of human behaviour models, all of them describing a specific domain, or modelling purpose, but to our knowledge there is no attempt to define model that can satisfy the requirements of activities from different domains, or even an attempt to identify the set of these requirements. We believe that human behaviour consists of a set of atomic actions that are the building blocks of more complex activities, and that they are uniform for the different types of activities. There already exists research in the field of activity recognition that tries to identify and recognize such activity primitives (e.g. [18, 5]). Based on this assumption, our research aims to define a set of atomic actions, that can be reused in different applications; and

Kristina Yordanova
MMIS, Computer Science, University of Rostock, Germany
e-mail: kristina.yordanova@uni-rostock.de

P. Novais et al. (Eds.): Ambient Intelligence - Software and Applications, AISC 92, pp. 247–251.
springerlink.com

to derive a uniform modelling approach that is applicable in different scenarios. We hope that the reusability of such approach will increase the functionality of assisting systems that perform intention recognition. In this paper, we describe the progress of our research in respect to the activity analysis and action constants definition; and the requirements for a uniform HBM language, derived so far, as well as the models that we have already investigated and the requirements they satisfy.

2 Related Work

When talking about activity recognition, probabilistic models are probably the first that come to mind. There is a lot of past and ongoing research on recognizing human activities and intentions with the help of HMM such as in [1, 2, 3].

A problem in using probabilistic models for activity recognition is the exponential state growth with the increase of possible actions. To solve this problem Burghardt et al. generates Markov models by utilizing partial order planning [4].

Another approach to intention recognition that is getting more popular in the recent years is based on statistical methods such as feature extraction or feature selection combined with different classifiers [6, 5].

Above we listed some methods used for activity recognition. However, in our research we are interested in the underlying models that can help us recognize a performed activity. Such methods or languages for HBM are CTML [7], ACT-R [8], GOMS [9], PMFserv [11], PECS Reference Model [12], Behaviour Design Patterns [13], Rasmussen's Human Behavior Model [14], Concurrent Task Trees [15].

Although the described modelling methods show encouraging results and are able to successfully model human behaviour and to identify activity primitives, there is no attempt to generalize these primitives or to derive a uniform modelling approach for scenarios with different contexts. Next we give our preliminary concepts and results in attempt to build such generalized HBM approach.

3 Atomic Action Constants

The first phase of the research aims at deriving a set of activity primitives that are valid for various activity domains. We believe the set of atomic constants can be represented by a small set of parameterisable action schemes, that are, for instance, representable as PDDL operators. At present we have datasets from the domains of outdoor activities, indoor activities, smart meeting room activities, behaviour of people with cognitive restrictions and elderly care activities. These datasets are analysed and the atomic action constants in them are derived. Analysing the elderly care dataset has resulted in 12 atomic action schemes such as *move* (see Fig. 1), that make up the 2847284 observed atomic actions. The process of defining atomic activities is done by decomposing the complex activities until the simplest actions are reached. The atomic activities derived so far from the elderly care activities dataset are *move, take, put, do, go, push, turn, cut, wait,* where every action is described in a PDDL notation. Fig. 1 shows the preconditions and effects for the action *move.*

```
:action move (?p - person ?o - object ?from ?to - place)
        :preconditon (and (at ?p ?from) (holds ?p ?o))
        :effect (and (at ?p ?to) (at ?o ?to) (not (at ?p ?from)) (not (at ?o ?from))
```

Fig. 1 Action *move* in PDDL notation

When a concrete use case is present the atomic activities are parameterised with specific objects, locations, etc. and sensor data is employed that is either independent from parameterisation, or can be estimated from concrete parameters. The atomic actions are also used for composing reusable complex activities.

4 Requirements for Human Behaviour Models

Parallel to the atomic actions definition, we apply requirements analysis to the activity datasets in order to identify the model requirements they have. Thus, we gather a set of requirements from different domains that is the base for the definition of a modelling approach able to catch the activities' dynamics. Let us consider human behaviour as a complex system consisting of a set of elements with relationships between them. To model such system, there are requirements that the modelling language has to meet. Applying requirements analysis to the datasets from Section 3, the following groups of requirements for uniform HBM language are derived:

- **for procedural modelling:** composition, hierarchy, sequences, loops, interleaving activities, choice, constraints, enabling, disabling, priority, independence, suspend, resume;
- **for parallel execution modelling:** parallelism, synchronisation;
- **for probabilistic modelling:** observation models, probabilities for action sequences, probable durations of activities;
- **for causal modelling:** preconditions, effects, relation to prior knowledge;
- **modelling purpose:** simulation, prediction, causal inference, state estimation, parameter estimation, detecting errors, unknown actions detection;

Based on this set of requirements, we now investigate different existing models and try to discover which of the requirements they satisfy. The reason for this is to establish which methods are suitable to be combined or further extended, in order to be used in a uniform HBM approach. The models we have investigated so far are as follows: PECS reference model [12], ACT-R [8], PMFserv [11], SOAR [16], GOMS [10], BDI [17], BDP [13], CPM-GOMS [9], CTT [15], and CTML [7]. Unfortunately, no model fulfills all the requirements, the most compatible being CTML because it satisfies most of the requirements with the exception of the cognitive psychology capabilities. Moreover, none of the models satisfies the conditions concerning sensor data handling and probabilistic parameterisation. In order to solve these problems, we believe that a method should be chosen, that allows the mapping into a probabilistic model such as Hidden Markov Model. Thus, procedural models, encoding prior knowledge about stereotypical action sequences, will be combined with declarative models, allowing an efficient representation of all causally valid action sequences in order to achieve a goal.

5 Open Problems

In Sections 3 and 4 we described the current state of the research. However, there are open issues that have to be resolved.

- *Are the atomic constants uniform for different domains?* The present set of atomic actions is derived from the elderly care activities dataset and is still not validated with other activity datasets.
- *When does an activity decomposition has reached the atomic activities level?* At present we assume that the atomic level is reached when the action can be used in different contexts and domains.
- *What are the relations between the different atomic constants?* So far we have only defined the atomic constants but not the relationships between them.
- *Is the set of HBM requirements complete?* Although the derived set of requirements is based on scenarios from different domains, we are not aware of an approach that can validate that this set is complete.
- *Not enough information on some of the methods.* Many of the investigated modelling methods are described in papers that do not provide sufficient information about the modelling mechanism.
- *What other HBM methods exist and are they more suitable for our purpose?* At present we have investigated several methods, but there is still the possibility that there exists a method that is more suitable for our purpose.

6 Conclusion and Future Work

In this paper we described a project that aims to define a uniform HBM approach for intention recognition in smart environments. As the research is still in early stages, the definition of sets of action constants and HBM requirements is still in progress. However, the preliminary results show that it is possible to derive uniform requirements and action constants sets. Additionally, some existing modelling approaches were investigated and the requirements they satisfy were identified to a degree.

The next steps of the research will be focused on completion and validation of the set of activity constants and modelling requirements, as well as on definition of the relationships between the different action constants and the choice of appropriate methods to be used and possibly combined in a unified HBM approach.

Acknowledgements. This work is supported by the DFG, research training group MuSAMA (grant no. GRK 1424/1).

References

1. Aarno, D., Kragic, D.: Motion Intention Recognition in Robot Assisted Applications. Elsevier, Amsterdam (2007)
2. van Kasteren, T., Noulas, A., Englebienne, G., Krose, B.: Accurate Activity Recognition in a Home Setting. In: Proceedings of the UbiComp Conference (2008)
3. Modayil, J., Bai, T., Kautz, H.: Improving the Recognition of Interleaved Activities. In: Proceedings of UbiComp (2008)
4. Burghardt, C., Kirste, T.: Synthesizing Probabilistic Models for Team-Assistance in Smart Meetings Rooms. In: Proceedings of the CSCW Conference (2008)
5. Albinali, F., Goodwin, M.S., Intille, S.S.: Recognizing Stereotypical Motor Movements in the Laboratory and Classroom: A Case Study with Children on the Autism Spectrum. In: Proceedings of the UbiComp Conference (2009)
6. Bulling, A., Ward, J.A., Gellersen, H., Troster, G.: Eye Movement Analysis for Activity Recognition. In: Proceedings of the UbiComp Conference (2009)
7. Wurdel, M., Sinnig, S., Forbrig, P.: CTML: Domain and Task Modeling for Collaborative Environments. Journal of Universal Computer Science (2009)
8. Foyle, D.C., Hooey, B.L., Byrne, M.D., Corker, K.M., Deutsch, S., Lebiere, C., Leiden, K., Wickens, C.D.: Human Performance Models of Pilot Behavior. In: Proceedings of the Human Factors and Ergonomics Society (2001)
9. Lee, S.M., Remington, R., Ravinder, U., Matessa, M.: Developing Human Performance Models Using Apex / CPM-GOMS for Agent-Based Modeling and Simulation. In: Proceedings of the Conference on Winter Simulation (2005)
10. Tonn-Eichstaedt, H.: Measuring Website Usability for Visually Impaired People - A Modified GOMS Analysis. In: Proceedings of the ASSETS Conference (2006)
11. Silverman, B.G., Johns, M., Cornwell, J., OBrien, K.: Human Behavior Models for Agents in Simulators and Games: Part I Enabling Science with PMFserv. Presence: Teleoperators and Virtual Environments (2006)
12. Schmidt, B., Schneider, B.: Agent-Based Modelling of Human Acting, Deciding and Behaviour – The Reference Model PECS. In: Proceedings of the European Simulation Multiconference (2004)
13. Taylor, G., Wray, R.E.: Behavior Design Patterns: Engineering Human Behavior Models. In: Proceedings of the Behavior Representation in Modeling and Simulation Conference (2004)
14. Wentink, M., Stassen, L.P.S., Alwayn, I., Hosman, R.J.A.W., Stassen, H.G.: Rasmussens Model of Human Behavior in Laparoscopy Training. Surgical Endoscopy (2003)
15. Giersich, M., Forbrig, P., Fuchs, G., Kirste, T., Reichart, D., Schumann, H.: Towards an Integrated Approach for Task Modeling and Human Behavior Recognition. In: Proceedings of Human-Computer Interaction Conference (2007)
16. Crossman, J.: Aspects and Soar: A Behavior Development Model. In: 27th Soar Workshop (2007)
17. Norling, E.: Capturing the Quake Player: Using a BDI Agent to Model Human Behaviour. In: Proceedings of the IEEE International Conference on Video and Signal Based Surveillance (2006)
18. Tolstikov, A., Biswas, J., Chen-Khong, T., Yap, P.: Eating activity primitives detection - a step towards ADL recognition. In: Proceedings of HealthCom Conference (2008)

Evolutionary Intelligence in Agent Modeling and Interoperability

Miguel Miranda, José Machado, António Abelha, José Neves, and João Neves

Abstract. A healthcare organization to be tuned with the users expectations, and to act according to its goals, must be accountable for the quality, cost, and overall care of the beneficiaries. In this paper we describe a model of clinical information designed to make health information systems properly interoperable and safely computable, based on an Evolutionary Intelligence approach that generates quantified scenarios from defective knowledge. The model is a response to a number of requirements, ranging from the semantic ones to the evaluation of software performance at runtime; it is among the biggest challenges in engineering nowadays.

Keywords: Multi-agent Systems, Healthcare, Intelligent Systems.

1 Introduction

Considering the course and the initial impetus on agent-based systems and agent-based methodologies for problem solving, it is understood that it is not available, yet, a real bunch of practical applications and implementations outside the realm of research in software analysis and development. However, agent-based systems are arriving at a stage of maturity, i.e., there are tools such as JADE, WADE and others frameworks that started a consensus that may help to add agent-based systems to the main stream of programming methodologies for problem solving.

When approaching a problem, any interpretation cannot be bonded, *a priori*, to a specific technology or methodology for problem solving. New systems must be

Miguel Miranda · José Machado · António Abelha · José Neves
Univeristy of Minho
Departmento de Informática
e-mail: {miranda,jmac,abelha,jneves}@di.uminho.pt

João Neves
Centro Hospitalar de Vila Nova de Gaia e Espinho EPE
e-mail: j_neves@hotmail.com

P. Novais et al. (Eds.): Ambient Intelligence - Software and Applications, AISC 92, pp. 253–257.
springerlink.com

oriented and adapted to better fit into the environment. Using multi-agent systems or agent-based models for problem solving presents a different way to set different courses in research methodology and practice, i.e., a system of methods used in a particular area of study or activity, that can be imbedded in an organized manner.

On the other hand, a Service Oriented Architecture (SOA) denotes, fundamentally, a collection of services, which communicate with one another. The communication process can involve either simple data going past or across or it could include two or more services negotiating some activity; where a service is to be understood as a function that is well-defined, self-contained, and does not depend on the context or state of other services. Connecting services is therefore paramount; indeed enterprise applications and software systems need to be interoperable in order to achieve seamless business across organizational boundaries. It may seem extremely similar to the MAS approach, however SOA is an abstraction not bound to one specific technology, once it can be based on web-services, agents or in any other framework following these basic rules. The same can be applied to ontologies for cross-organizational communication, which have been already integrated in existing agent development frameworks. Indeed, new business models also call for innovative approaches to customers, involving collaboration across different organizations and domains and therefore need cross-organizational communication.

2 A Multi-Agent Environment for Distributed Interoperability in Healthcare

In order to enable increased levels of interoperability, a whole environment centered in agent-based systems was developed to take advantage of the agent based methodology for problem solving. As it is visible in Figure 1, the applicational environment can be divided according to the following tiers of abstraction: Multi-Agent Distributed Interoperability Platform (MADIP); Web-based Interoperability Platform (WIP); Hibernate and the Relational Database Management System.

The external service providers depicted in Figure MAEDIH include different information systems such as the Radiological Information System, Laboratory Information System, Hospital Emergency Rooms Information Systems and other back-end and front-end to external applications. This environment went in production at the Centro Hospitalar do Porto, EPE, a major healthcare facility in the north of Portugal, as an element of the agency AIDA (Figure 2) (i.e., an Agency for the Integration, Diffusion and Archive of Information), that is in charge of the interoperability among all the services present in a healthcare environment, here depicted as a Healthcare Information System (HIS). A HIS can be defined as an abstract information system for data processing within an healthcare institution. It is therefore the consorted and integrated effort of the different, tentatives, heterogeneous information systems inside the healthcare institution, which collects, processes, reports and use information and knowledge within this unique environment, i.e., that influences the existing management policies, health programs, training, research and medical practice within the institution [5] [6].

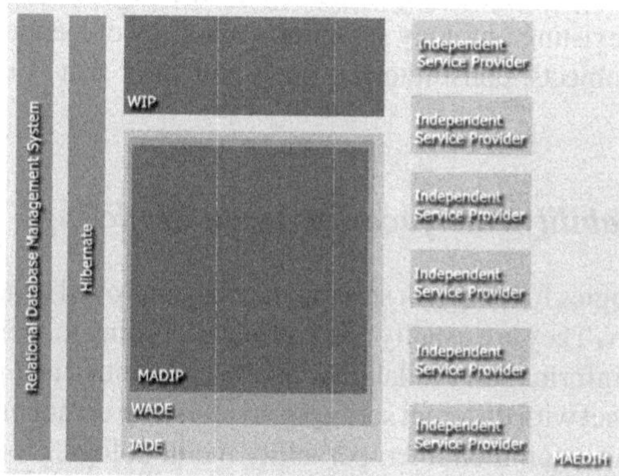

Fig. 1 Core tiers of the integration environment

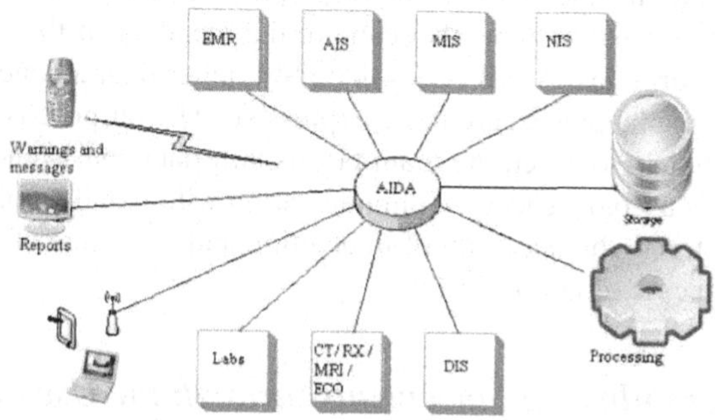

Fig. 2 AIDA platform

2.1 Core Architectural Dependencies

Regarding the MADIP architecture, that is a multi-agent system developed under the Workflows and Agents Development Environment (WADE) and the Java Agent DEvelopment Framework (JADE). This choice was taken considering that JADE is an Open-Source JAVA based solution which offers some advantages, namely the production environment tests, a continuous improvement in any new release, and the demonstrated stability that assures the support and capability to handle all the events that may occur in a production environment. Furthermore, the fact that the framework is fully compliant with the specification of the Foundation for Intelligent Physical Agents (FIPA) using the advised architecture and making available several agent communication codecs such as FIPA Semantic Language (SL) and XML, enables one to extend the existing framework to a particular setting, without "reinventing the wheel" [1]. Complementary to the Agent Communication Language (ACL), the existing ontology server and ontology embedded communication trough plain Java objects, constitutes an important asset that is made available by JADE.

2.2 Interoperability Interfaces in Agent Modeling

Diverse methodologies for problem solving may be employed to achieve integration and interoperability. They are usually comparable according to their specificities and the necessities on information availability, quality, quantities and processing time.

In order to interact with different service providers that do not use ACL with FIPA compliance, the most common interoperability methodologies for problem solving are put into effect through configurable agents, using the methods: Health Level Seven (HL7) *via* TCP socket; HL7 *via* SOAP and Web-Services; AIDAs XML *via* SOAP and Web-Services and Legacy System via TCP socket. Using plain Java objects as mappings for the XML and HL7 structures being provided, it is possible to associate them to diverse medical ontologies, here given in terms of Knowledge Bases (KBs), i.e., formal systems that capture the meaning of the adopted vocabulary via logical formulas. A KB is considerably richer than a conceptual schema since the underlying languages are more expressive. The purpose is not simply retrieval but reasoning. However, the main task is still data consistency. Formally, a knowledge base is a Logic Theory. Communication within and among agent-based platforms, considering the necessity of interaction and exchange of information, is without a doubt a case of interoperability.

2.3 Interoperability Enhancements through the Use of Ontologies

When interoperability is accounted in terms of the nature of being, i.e., in terms of an or more ontologies, it has the potential to improve the flow of information and knowledge exchange inter peers. Tis is a good or obvious cause to use the WADE/JADE based ontologies to manage content expressions as Java objects across domains and applications [4]. Although they are handled as Java objects, they are encoded and decoded into messages in a standard FIPA format. The ontology engine in JADE is made on a name, a running ontology that is susceptible of a particular figure or combination and a set of element schemas which may contain 3 (three) basic interfaces, namely Agent Action, Predicate Extensions and Concepts[2], i.e., these element schemas are objects describing the structure, semantics and relationships among the Agent Action, Predicate Extensions and Concepts that hold within a given ontology, which may share vocabulary among the different agents using it. Rather than the Ontology, BasicOntology and ACLOntology, the BeanOntology can be used to map JavaBeans to an ontology using a "convention over configuration" perspective [3]. By this mean JavaBeans annotations are used to extend the objects that constitute the ontology. The Agent Action denotes a content expression of a direct request among agents to perform a given task. This action request will contain the information regarding what action and to what objects it is to be applied. For this reason all requests received from any service provider were transformed through object mappings into an Agent Action that was communicated to the agents involved in the process. Whenever an agent asks for the truth value of a given sentence, the sentence can be defined as a class that implements a Predicate Extension[2].

In the proposed ontology, the definition for interoperability, Agent Actions and Predicates very often holding Concepts. A class that implements a Concept interface is composed of several objects implicitly conjuncted. Existing mapping using Hibernate are also compatible to be mapped within these concept classes making it easy to exchange information between agents while the content is completely normalized and congruent with the storage schema. The use of ontologies over serialized objects allows one not only have interoperability with other agent-based systems, but, to a some extent, to have interoperability with further non agent-based systems.

3 Conclusion

This environment was tested in the first place and put into production in an ambulatory surgery centre as a part of a huge healthcare facility. The need to interoperate was felt at almost all levels of decision as defined by the business logic, which is not part of this document. We argue that the starting point of a successful model must be an ontological analysis of the process of clinical care delivery, seen as a scientific problem-solving process. The combination of Artificial Intelligence techniques such as the symbolic and evolutionary systems allows us to combine the advantages of each of these approaches, in particular for solving complex problems. These techniques are based on the collective adaptation and learning ability of individuals.

References

1. Bellifemine, F., Caire, G., Trucco, T., Rimassa, G.: Jade programmer's guide. Tech. rep., Telecom Italia S.p.A. (2010)
2. Caire, G., Cabanillas, D.: Application-defined content languages and ontologies. Tech. rep., Telecom Italia S.p.A. (2010)
3. Cancedda, P., Caire, G.: Ade tutorial creating ontologies by means of creating ontologies by means of the bean-ontology class. Tech. rep., Telecom Italia S.p.A. (2008)
4. Ghosh, K., Natarajan, S., Srinivasan, R.: Decision fusion in distributed multi agent process supervisory system. In: International Conference on Networking, Sensing and Control (ICNSC 2010), pp. 183–188 (2010), `http://ieeexplore.ieee.org/stamp/stamp.jsp?arnumber=5461510`
5. Kirsh, W. (ed.): Encyclopedia of Public Health, vol. 1. Springer, Heidelberg (2008)
6. Miranda, M., Pontes, G., Gonçalves, P., Peixoto, H., Santos, M., Abelha, A., Machado, J.: Modelling Intelligent Behaviours in Multi-agent Based HL7 Services. SCI, vol. 317, pp. 95–106. Springer, Heidelberg (2010)

Author Index

Álvarez, Laura 93

Abelha, António 253
Almeida, Ana 133
Alonso, Alonso A. 17
Alonso, Ricardo S. 173
Alonso, Vidal 93
Amorim, Tiago 241
Anacleto, Ricardo 133
Ascanio, Juan R. 33
Augusto, J.C. 109
Augusto, Juan Carlos 1, 197

Bajo, Javier 93, 173
Berbers, Yolande 41
Black, Norman 181
Bonino, Dario 213
Botia, Juan A. 189

Callaghan, Victor 59
Campuzano, Francisco 189
Carneiro, Davide 25
Carneiro, João 69
Carrera, Albano 17
Carswell, W. 109
Cecchi, David Olivieri 101
Chouvarda, Ioanna 141
Conde, Iván Gómez 101
Corno, Fulvio 9, 213
Correia, Nuno 241

Davies, Marc 59
de Almeida, Ana 51
de la Rosa Steinz, Ramón 17
de Paz, Juan F. 77
De Russis, Luigi 213
del Val, Lara 17

El Fallah Seghrouchni, Amal 165

Faria, Luiz 219
Fdez-Riverola, Florentino 85
Felisberto, Filipe 85
Fernández, Carlos 227
Fernández, Javier Martínez 1
Fernandez-Montes, Alejandro 205
Figueiredo, Lino 133
Figueiredo, Marisa 51
Florea, Adina Magda 165
Fraile, Juan A. 77
Freitas, Carlos 69
Fuentes, Daniel 205

Gómez, María I. Jiménez 17
Galindo, C. 125
García, Oscar 173
García-Herrero, Jesús 117
García-Molina, Jesús J. 41
Garcia-Sola, Alberto 189
Garcia-Valverde, Teresa 189
Giráldez, Ignacio 33
Gomes, Luís 219
Gonzalez, J. 125
Gonzalez-Abril, Luis 205
Griol, David 117
Guevara, Fabio 173

Hawkey, Jared 241
Hoyos, José R. 41

Jeffers, P. 109

Koutkias, Vassilis G. 141

Laranjeira, João 69
Lawless-Reljic, Sabine 59
Leistikow, René 149
Luz, Nuno 133

Machado, José 253
Madrid, Natividad Martínez 1
Maglaveras, Nicos 141
Marcelino, Isabel 85
Marques, Albino 219
Marreiros, Goreti 69
Martín, Patricia 93
McCullagh, Paul J. 181, 197
Meneu, Mª Teresa 227
Miranda, Miguel 253
Molina, José M. 117
Moreira, Nuno 85
Mulvenna, M. 109

Navarro, Mª Amparo 227
Neves, João 253
Neves, José 25, 253
Novais, Paulo 25, 133
Nugent, Chris 181

Olaru, Andrei 165
Oliveira, Sofia 241
Ortega, Juan A. 205

Páez, Diego Gachet 33
Pérez-Lancho, Belén 77
Pereira, António 85
Preuveneers, Davy 41

Ramos, Carlos 69, 219
Ribeiro, Bernardete 51
Rodríguez, Sara 77
Rodríguez, Ángel Orosa 101
Rodrigues, Armanda 241
Rubio, Margarita 33
Ruiz-Sarmiento, J.R. 125

Sánchez, Ana Belén 227
Sánchez, Miguel 93, 173
Sanaullah, Muhammad 9
Sancho, David 173
Seepold, Ralf 1
Sharpanskykh, Alexei 233
Silva, António 219
Sobrino, Xosé Antón Vila 101

Triantafyllidis, Andreas K. 141

Vale, Zita 219
Vatavu, Radu-Daniel 157

Wang, H. 109
Wang, Minjuan 59

Yordanova, Kristina 247

Zhang, Shumei 181
Zheng, H. 109
Zheng, Huiru 181
Zia, Kashif 233